The Maintenance Mechanic's/ Machinist's Toolbox Manual

John D. Bies

ARCO

New York

Art figures 1-2, 1-3, 1-4, 1-5, 1-6, 1-7, 1-8, 1-9, 1-10, 1-11, 1-12, 1-15, 1-16, 1-18, 1-19, 1-20, 1-21, 1-22, 1-24, 1-25, 1-26, 1-27, 1-28, 1-29, 1-30, 1-31, 1-32, 1-33, 1-34, 1-36, and 1-37 appear courtesy of Stanley Tools, a division of The Stanley Works.

ARCO

Simon & Schuster, Inc.
Gulf+Western Building
One Gulf+Western Plaza
New York, NY 10023

DISTRIBUTED BY PRENTICE HALL TRADE

Manufactured in the United States of America

1 2 3 4 5 6 7 8 9 10

Library of Congress Cataloging-in-Publication Data

Bies, John D.,
The maintenance mechanic's/machinist's toolbox manual.
Includes index.
1. Tools—Handbooks, manuals, etc.
2. Maintenance—Handbooks, manuals, etc.
I. In Plain English, Inc. II. Title.
TJ1195.B54 1989 621.9 88-35111
ISBN 0-13-545666-5

CONTENTS

Preface

The trade area of machine and maintenance mechanic is perhaps one of the most technically diversified. It requires knowledge and skills in carpentry, electricity, fluid power, machining, and other fields. In light of the breadth of competencies required of these tradespeople and apprentices, much thought and research went into the organization of this book.

The Maintenance Mechanic's/Machinist's Toolbox Manual was written to serve as a ready reference for both tradespeople and apprentices. Throughout the book, photographs and line drawings are used to illustrate tools and procedures. Each section and chapter is organized to make problems and jobs encountered in the field easy to solve and execute. Both the alphabetical format of the chapters and the index allow quick access to specific information.

The first part of the book deals with the tools of the trade. Here, general maintenance tools, measuring and testing tools, and power tools are discussed in terms of their uses, proper handling, application, and care.

The second part of the book covers preventive maintenance and troubleshooting. Four chapters address typical problems encountered with building and plant facilities, electrical equipment, mechanical equipment, and service equipment.

Part Three deals with the ABC's of the trade. Information on corrosion, lubrication, and welding is included, along with specific maintenance servicing information for buildings and facilities, electric equipment, mechanical equipment, and service equipment. Again, a sequential, step-by-step format is employed whenever possible.

The fourth part of the book deals with trade fundamentals, including trade mathematics, measurement and conversion, and blueprint reading. Several appendices and a comprehensive glossary complete the book.

I would like to make special note of the significant contributions of John Xerri at Henry Ford Community College in Dearborn, Michigan, who served as a reviewer for this book. His professional and technical expertise proved to be an invaluable asset.

John D. Bies

PART ONE

TOOLS OF THE TRADE

1

Hand Tools

The starting point of every maintenance program is the purchase of tools. It has been said that true artisans are recognized by the way they handle their tools. The way tools are handled includes not only how they are used but also how they are treated when not in use.

Every machinist and maintenance mechanic should become familiar with the various types of tools they will need, how to use them, and how to take care of thèm. Tools that are frequently used, are versatile, and require low maintenance should be selected first. Table 1-1 lists the hand tools most frequently used by maintenance mechanics.

TABLE 1–1
Mechanics' Basic Hand Tools

Number Needed	Tool Description
1	Tool box approximately 32 × 8 × 8 inches with lid, hasp, and lock
1	Pocket rule
1	Tape measure, 50 feet
1	Level
1	Crosscut saw, 2– 8 point

TABLE 1–1 (*continued*)

Number Needed	Tool Description
1	Ripsaw, 6 point
2	Hammers, curved claw and peen
1	Nail bar, 24 or 30 inches
1	Pliers, heavy-duty wire cutters
1	Allen wrench set
1	Adjustable wrench, 10- or 12-inch
1	Socket and open-end wrench set (U.S. Customary and Metric)
1	Tube cutter set
1	Screwdriver, slot
1	Screwdriver, Phillips
2–3	Pencils
1	Scribe
1	Dividers
1	Nailset
1	Punch set
1	Tin snips
1	Hack saw
1	Coping saw
1	Twist drill (drills: 1/16 inch to ¼ inch)
1	Brace bits: ¼ to 1½ inch

A toolbox provides mechanics with flexibility on the job. In effect, it becomes a portable workshop. Because one is limited in the number and size of tools that can be transported, only essential and frequently used tools should be carried. Specialized tools may be added or removed as required.

Much maintenance and repair work done by mechanics is executed at a workbench. The primary requirements of this work station are that the bench be strong, rigid, and large enough that work can be performed easily. The correct height is important and varies according to the type of equipment to be used.

It is not necessary to have a great range of tools and equipment for benchwork. The equipment may include 360- and 180-cycle outlets, an air supply, vises, hammers, saws, files, sockets, closed and open-ended wrenches, a bench punch, and gages and other measuring equipment.

BORING TOOLS

A number of different types of tools are used to bore holes in material and to perform related operations. All augers and bits held by a brace or hand drill are considered boring tools.

BRACE AND BITS

The common bit brace, auger bits, and several specialized tools fall into this classification. They are used primarily to drill holes in soft material such as wood and plaster board.

COMMON TYPES OF BITS

Auger Bit. An auger bit is a general term for bits used on a bit brace. The bit is used primarily to bore holes in wood and other soft materials (See Figure 1-1.) Common auger bit sizes range from 1/4 inch to 1 inch. Larger bit set sizes are also available, though expansive and lock set bits are used for drilling large holes.

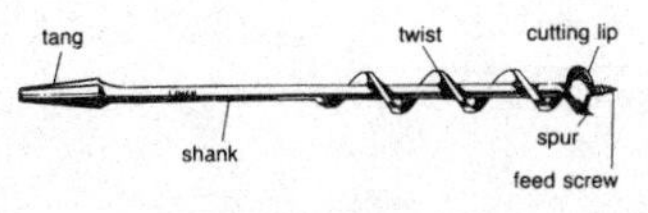

Figure 1-1. Auger bit.

Bit Gage. This is a gage that is attached to a drill bit for boring a hole to a specific depth. It functions as a stop, or a go–no-go, gage, so that when the required depth is reached, it prevents the drill from penetrating any further. (See Figure 1-2.)

Figure 1-2. Bit gage.

Countersink Bit. Unlike most bits, a countersink bit cannot actually drill a hole. Countersink bits are used to form a conical shape or countersink. They are available in sizes up to 3/4 inch. (See Figure 1-3.)

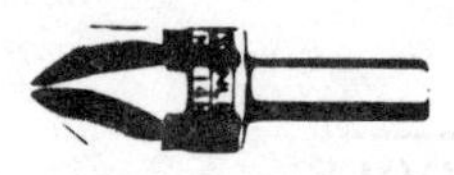

Figure 1-3. Countersink bit.

Extension Bit. When additional length is needed for a bit, an extension bit is used. Extension bits come in common lengths ranging from 18 inches to 24 inches. They are used to bore holes in thick stock such as timber and in form construction where two holes must be aligned over extensive distances. (See Figure 1-4.)

Figure 1-4. Extension bit.

Forstner Bit. A Forstner bit is used primarily to bore shallow holes part way through stock. The bit cuts with two lips and a circular rim. There is no screw point available to center it during drilling.

Screwdriver Bits. These bits are extremely useful for driving large screws into heavy stock or hardware. They are also used for removing old screws that are difficult to remove with a regular screwdriver. Screwdriver bits are available with either a straight or Phillips tip.

TYPES OF BRACES

Corner Brace. This type of brace is used primarily for boring operations in corners and against walls.

Ratchet Bit Brace. A typical ratchet bit brace has a 10-inch sweep, which is the diameter of the half circle made when turning the handle. With the ratchet engaged, the bit brace handle will apply forward pressure on the bit when turned through part of a circle. The handle can be rapidly rotated in the opposite direction to the starting point and turned again to continue the forward motion of the bit. This feature allows mechanics to work close in corners or to walls. A high quality ratchet bit brace has an adjustment that allows it to turn either counterclockwise or clockwise. (See Figure 1-5.)

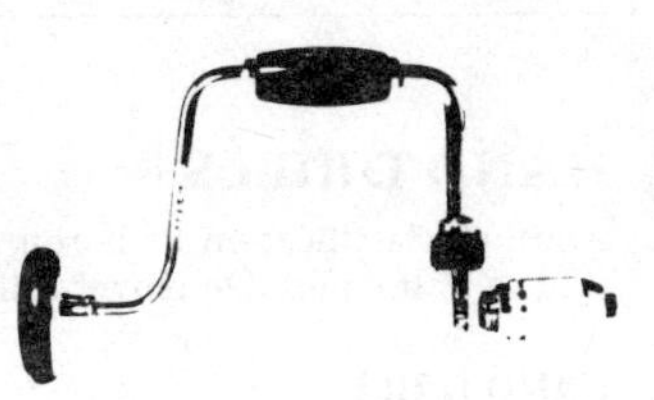

Figure 1-5. Ratchet bit brace.

PROPER USE OF BRACE AND BITS

- Always keep your attention focused on the work.
- Maintain bits so that they are sharp.
- Select the correct type and size bit for the job.
- Once the bit has been inserted into the brace, make sure that it has been securely tightened.
- For accuracy, make a hole with an awl so that the bit can be started on center.
- Keep the brace straight and at the correct angle to the workpiece (usually at a right angle). Never twist the brace to one side or the other, and maintain a steady pressure.
- When possible, do not bore all the way through the material. Complete the boring operation from the other side.
- Do not drive the bit any deeper than its twist; otherwise clogging and overheating will occur.
- To remove the bit from a hole, reverse the direction of the bit. Do not pull it straight back without turning the bit.

- After use, always clean bits with a rag and light coating of oil.
- Maintain the brace by periodically lubricating and cleaning it.
- Store braces and bits separately and in a suitable place in the shop or your toolbox.

HAND DRILLS

Another classification of boring tools is hand drills, which are similar to the portable power drill.

HAND DRILL

The basic tool used to bore holes from 1/4 to 3/8 inch in diameter is the hand drill. It operates when the worker turns the handle, which is geared to a chuck. (See Figure 1-6.)

Figure 1-6. Hand drill.

BREAST DRILL

The breast drill is similar in appearance and operation to the hand drill, except that it has a breast plate against which you can push your chest or shoulder for greater pressure. This tool is considered most useful in heavy-duty work and in drilling larger holes.

PUSH DRILL

The push drill is used for drilling small holes and installing small hardware. A semiautomatic tool, it can be operated with one hand. When the handle is pushed in, it rotates the drill. As pressure is released, an internal spring causes the drill to return to its original position. A set of bits or drill points, stored in the handle of the push drill, range in size from 1/16 inch to 1/64 inch. (See Figure 1-7.)

Figure 1-7. Push drill.

TWIST DRILL

The twist drill is the bit used to bore holes in material such as wood, metal, and plastics. The angle of the drill point depends on whether the material to be drilled is metal or wood. Carbon steel twist drills used for boring wood and other soft materials should not be used for drilling holes in hard metals. A twist drill used on hard metals will have either high speed (HS) or high speed steel (HSS) stamped on its shank. A drill that has no letter markings on the shank should be used for boring soft materials.

HAMMERS AND PERCUSSION TOOLS

This group of hand tools is used to drive objects, strike other tools, or form material. These tools are used frequently in benchwork as well as at the work site.

COMMON HAMMERING TOOLS

There are a number of different hammering tools available. The hammering tools presented here do not represent a complete survey of products available, but are the ones commonly used within the trade. (See Figure 1-8.)

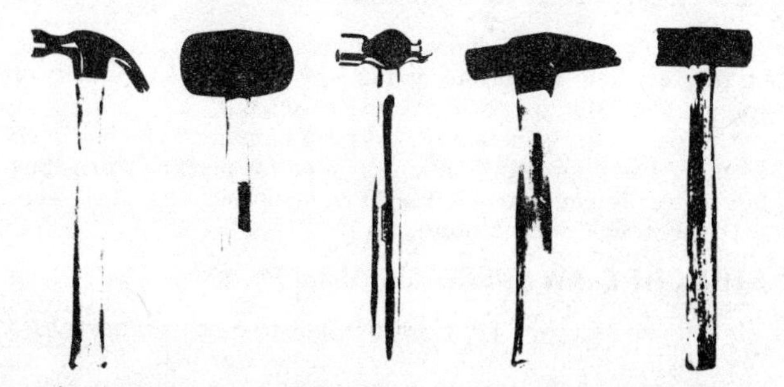

Figure 1-8. Hammers: curved claw, mallet, ball peen, straight claw, and sledge.

CURVED CLAW HAMMER

The most often used and recognized hammer is the curved claw hammer. Used for all types of carpentry work, it should be about 16 ounces for general purpose work. The head of the hammer should be made of high quality steel to withstand marring and chipping. The handle can be made of wood or metal but must be able to absorb some of the shock of the blow.

DRYWALL HAMMER

A hammer used for installing drywall and gypsum panels is the drywall hammer. It has a rounded convex face that "dimples," or indents, the drywall surface without breaking the paper cover.

FLOORING HAMMER

The flooring hammer is primarily used for laying tongue-and-groove hardwood floors. However, this type of hammer is often considered a specialized tool since a simple claw hammer can be used for floor-laying.

MALLET

A mallet is usually made of wood, rubber, or leather, and used for driving other tools. Mallets are often used for installing and setting up equipment and machinery. There are various weights and handle lengths available for different job requirements.

PEEN (MACHINIST'S) HAMMERS

A group of hammers known as peen, or *machinist's*, hammers can be of great help to the mechanic. Peens are used to indent or compress metal in order to expand or stretch it.

There are three general types of peen hammers. The ball peen hammer has a spherical end and is used for peening or riveting operations. The other two—straight and cross peen hammers—are used for straight indentations.

STRAIGHT CLAW (RIPPING) HAMMER

Also known as a *ripping hammer*, the straight claw hammer is used for rough carpentry work. Somewhat heavier than the curved claw hammer, it is often used to split pieces of wood as well as to drive and pull nails. The recommended weight for straight claw hammers is about 20 ounces.

SLEDGE HAMMER

Sledge hammers weigh between 2 and 20 pounds. They are used to drive stakes and batterboards and to break concrete and masonry. A lighter 2-pound sledge hammer is commonly used in timber construction when the lumber is 3 or more inches thick.

PROPER USE OF HAMMERS

- Make sure that the handle is securely fastened and free of splinters or any other defect.
- Be sure the face of the hammer is clean, not chipped, split, or mushroomed. If burrs are present, they should be ground off.
- Grip the hammer at the end of the handle. Use a light blow to set the nail and establish aim. Strike the nail squarely.
- Never strike with the side or "cheek" of the hammer.
- Do not strike hardened steel surfaces with a hammer. If they must be struck for setting or installation, use a mallet.
- Use claw hammers for pulling nails. Don't use hammers as wedges or pries.
- Store hammers and mallets in a proper location in the shop and toolbox.

PERCUSSION TOOLS

Several types of percussion tools are used by the maintenance mechanic. These tools not only save time but also make jobs easier.

MECHANICAL STAPLER (GUN TACKER)

Sometimes called a *gun tacker,* the mechanical stapler is used in a number of different jobs including tacking insulation, wall planking, and installing ceiling tile and metal lath. Staples used in these tools come in a variety of sizes up to 9/16 inch.

NAILING MACHINES

Manual nailing machines are used for a variety of jobs, including installing flooring underlayment and roofing material. A mallet is used to strike a plunger knob that ejects staples up to 1 3/4-inch long.

ROOFING HAMMER

Roofing hammers are specialized machines used to drive staples into asphalt or fiberglass strips of shingles. These hammers use staples that are 1-inch wide with 3/4-inch legs and are made out of 16-gauge wire.

LAYOUT AND BASIC MEASURING TOOLS

The efficiency and accuracy of the work performed by maintenance mechanics depends not only on their skills and training but also on their ability to measure and lay out work accurately.

BASIC MEASURING TOOLS

There are several basic measuring tools that the mechanic should have. Presented here is a description of those used for layout work. (For further information about measuring and testing equipment, see Chapter 2.)

BASIC RULES

The rule is the basic measuring tool used, and it is essential to the mechanic's trade. From it, a wide variety of other measuring devices have been developed to meet specific requirements. For example, hook rules are used to measure through holes in components such as gears and pulleys, and circumference rules are used to measure diameters and determine circumference by reading the scale.

There are two broad categories of rules that mechanics should know about: customary and metric rules. Customary rules use the inch–foot system of measure. Most common customary rules have four sets of graduations: eighths and sixteenths of an inch on one side, and thirty-seconds and sixty-fourths of an inch on the other side. Metric rules have divisions based on systems of ten. It is smart to purchase a metric rule marked with numbered lines in millimeters (mm). Metric rules divided into 1/2 mm are also available; they allow for even greater accuracy.

In addition to the common rule, there are several special types of rules used in maintenance work. These are the pull-push rule, the steel tape, and the zigzag rule.

Push-Pull Rule. A push-pull rule is a metal rule with graduations on the upper and lower edges of the rule. Push-pull rules come in standard lengths of 6, 8, 10, and 12 feet. Most of these rules are marked every 16 inches to help in stud layout. (See Figure 1-9.)

Figure 1-9. Push-pull rule.

Steel Tape. An important layout tool used for facility maintenance and machine installation is the steel tape. It is used to measure wall, room, and rafter lengths. The most common length used is a 50-foot steel tape.

Zigzag Rule. The zigzag rule is primarily used for carpentry and construction work. Be careful not to damage or break the blades when opening and closing a zigzag rule.

Figure 1-10. Zigzag extension rule.

The *extension rule* is a type of zigzag rule used to measure the inside of frames, such as window and door frames. It is used by opening the rule to within 6 inches or less of the opening and extending a sliding measure the remaining distance. (See Figure 1-10.)

SQUARES

Squares have one right angle and two straight edges for measuring and marking. (See Figure 1-11.) They can be used to determine the accuracy of surfaces that are supposed to be 90° to each other, lay out parallel lines, and set up workpieces in machinery.

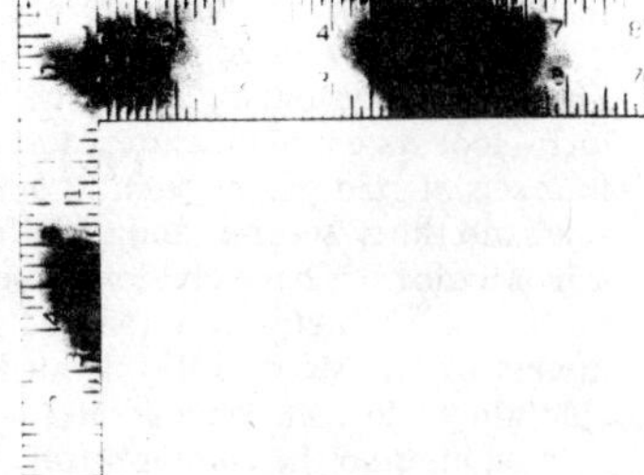

Figure 1-11. Steel square.

Combination Squares. Also known as a combination set, a combination square is made up of three separate tool heads—a center head, a protractor head, and a square head. It also has a graduated, hardened steel blade with a sliding tri-square head. Hence, a combination square can be used as a rule, square, miter for various angles, depth gage, height gage, and level. (See Figure 1-12.)

Figure 1-12. Combination square.

Framing Squares. As its name implies, a framing square is used in various framing operations, such as the framing of a building, the spacing of studs, and the framing of doors and windows. The standard framing square has a 24-inch blade 2 inches wide, and a 16-inch long tongue 1 1/2 inches wide. The blade and tongue are perpendicular to each other. Along these surfaces are division marks that are frequently used in carpentry work. When selecting a framing square, it is advisable to get one that has useful tables stamped on it, such as octagon scale, brace measure, and rafter-framing tables (see Figure 1-13).

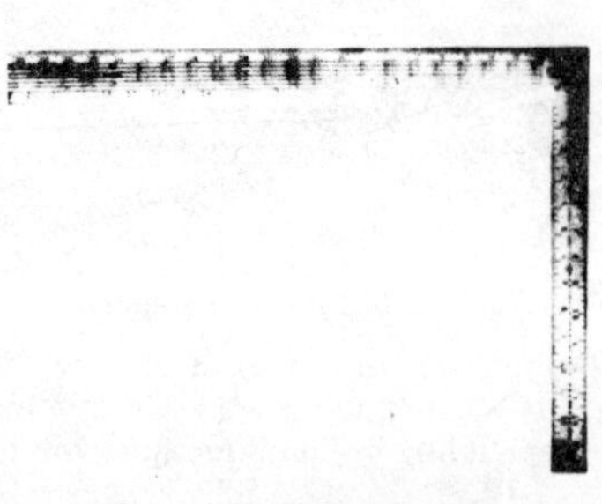

Figure 1-13. Framing square.

Try Squares. A try square has a blade from 6 to 12 inches long with a handle 4 3/8 inches to 8 inches long. In many shops, the try square has been replaced by the combination set.

BASIC LAYOUT TOOLS

Layout tools are used in the installation and building of equipment, machinery, and components.

CALIPERS

In layout operations, calipers are used to transfer measurements from one location to another. Calipers have two legs that can be opened and closed to the desired measure. (See Figure 1-14.) With these tools, it is possible to make measurements accurate to within 1/64 inch (0.40 mm). Generally, there are three types of calipers used by mechanics: hermaphrodite, inside, and outside calipers.

Figure 1-14. Pocket calipers.

Hermaphrodite Calipers. The hermaphrodite caliper has one pointed leg and one hooked leg. It is used to locate centers of cylindrical pieces and to scribe lines parallel to the sides of the objects. The tool is set by placing the hooked leg along the end of the rule and adjusting the pointed end to the desired setting.

Inside Calipers. Inside calipers are used to make internal measurements of cylindrical objects. Holes are measured by setting the caliper to the approximate size of the hole by inserting the legs into the opening.

Outside Calipers. External measurements are made with outside calipers. External measures are made by adjusting the legs of the calipers along the outside of the material or stock.

CHALK LINE

The chalk line and reel is used to mark a straight guide line on work. A chalked string is held taut and close to the work. It is then "snapped" so that it leaves a chalk line on the work. (See Figure 1-15.)

DIVIDERS

Dividers have two straight legs that are pointed at the end. They are used like a compass to lay out regular curves as well as to transfer measurements. A divider is set by placing one point of a leg on a rule marking and opening the other leg to the desired dimension. (See Figure 1-16.)

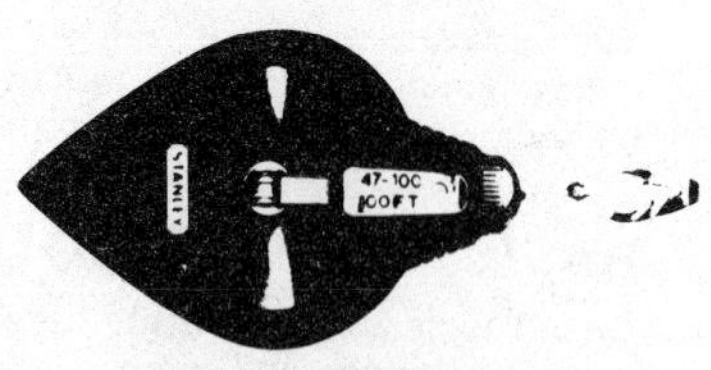

Figure 1-15. Chalk line.

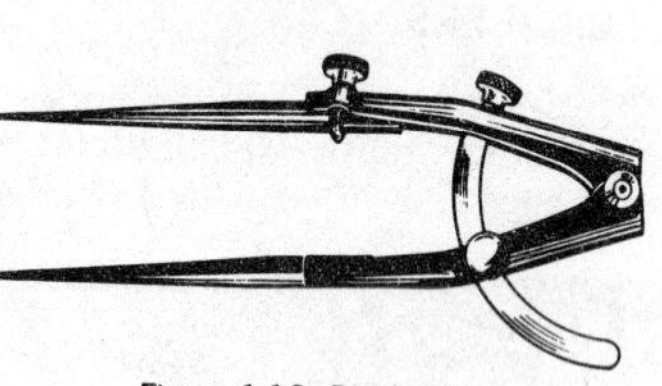

Figure 1-16. Dividers.

LEVEL

The level is used to level members of a building, machine, or equipment. The most common size is the 24- or 28-inch level. Some levels can be adjusted so that the glass vial containing the bubble can be reset as needed. To assure accuracy when using this tool, take two readings by reversing the ends. (See Figure 1-17.)

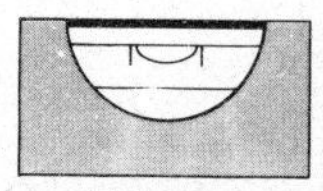

Figure 1-17. Reading a level.

PLUMB BOB

The plumb bob is used primarily to assure that an object is vertical, and it is usually used in combination with a level. The bob is suspended on a string along the height of the part or machine. Measurement is then taken to make sure that the object is parallel to the string. Plumb bobs are also used in surveying work to mark and set points. (See Figure 1-18.)

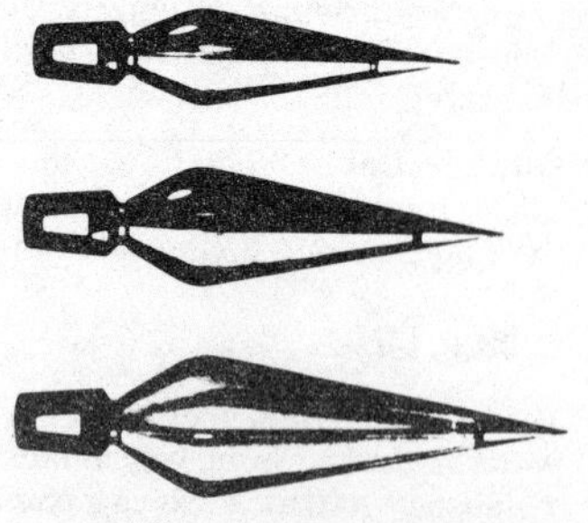

Figure 1-18. Plumb bobs.

PUNCHES

Punches are used to locate points, such as the center of circles, or to position tools. Two basic types of punches are used in layout work: prick punches and center punches.

Prick Punch. Used primarily in metal working, a prick punch is a layout tool used to mark the location of holes after lines have been scribed on the workpiece. (See Figure 1-19.)

Figure 1-19. Prick punch.

Center Punch. A center punch is similar to a prick punch except that its point is ground to an angle of 90°. It is used primarily to enlarge prick punch markings so that drill bits can be started more accurately and with greater ease.

SCRATCH AWL

The scratch awl is a layout tool that is used in carpentry work for locating positions and starting screws and other fasteners. The blade of the awl comes in sizes from 2 3/4 to 3 1/2 inches. (See Figure 1-20.)

Figure 1-20. Scratch awl.

STRAIGHTEDGE

Straightedges are used to draw or scribe straight lines. The size of the straightedge used depends on the specific need. Mechanics often make their own straightedges out of material that will not warp.

SHAVING AND SAWING TOOLS

Used on metal, wood, and other materials, shaving and sawing tools are important for accomplishing a number of repair and installation jobs.

SHAVING TOOLS

Shaving tools, which include a variety of chisels, files, and planes, are defined by the way they remove material.

COLD CHISELS

A cold chisel is a metalworking tool used to shear, cut, and chip cold metal. It is made of tool steel that is octagonally shaped and is sometimes referred to as chisel steel. One end is hardened so that it can withstand the blows of a hammer; the other end is ground to form the cutting edge.

PROPER USE OF A COLD CHISEL

- Keep your attention on the work at all times.
- Angle the chisel so that the chip cut will not be too deep.
- Strike the head of the chisel with a firm solid blow.
- After each blow, reset the chisel.
- As soon as a mushroom starts to form on the head of the chisel, grind it off at an angle of 60° to 70°.

There are four types of cold chisels commonly used in metalwork, and they come in a variety of sizes. (See Figure 1-21.)

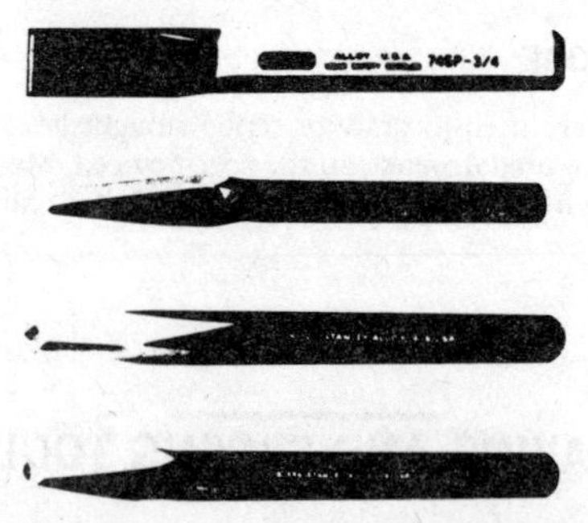

Figure 1-21. Cold chisels: flat cold chisel, cape chisel, diamond point chisel, and half round nose chisel.

Cape Chisel. As its name implies, the cape chisel is made to produce a cape or flare. Its cutting edge is narrow in relation to its body, so that it will not stick in a groove or slot. It is recommended for cutting narrow slots or keyways.

Diamond Point Chisel. The diamond point chisel is used for chipping V-shaped grooves and corners. It gets its name from a diamond-shaped cutting surface that resembles the facet corner on a diamond.

Flat Cold Chisel. A flat cold chisel is used for general metal chipping and cutting jobs. It is commonly used for the removal of rusted rivets, bolt heads, and nuts.

Round Nose Chisel. In a round nose chisel the cutting edge is ground to an elliptical shape. These chisels are used to cut concave surfaces, such as those needed for oil grooves.

WOOD CHISELS

Chisels that are used to cut away wood are constructed differently from cold chisels. The blade in a wood chisel is fastened to a wooden or plastic handle and ranges from 1/8 inch to 2 inches in width. (See Figure 1-22.) An exception to this is the flooring chisel, which is an all-metal chisel used on wood containing nails and other obstructions. In many cases, mechanics purchase wood chisels in a kit with three sizes: 3/8, 3/4, and 1 1/2 inch.

Figure 1-22. Wood chisel.

FILES

Filing is considered either a finishing or semifinishing operation. There are two broad methods of using a file. The first is straight filing, which is accomplished by pushing the file along the length of its blade—either parallel to or at a slight angle to the workpiece. The other method is draw filing in which the file is moved in short strokes usually perpendicular to the workpiece; this technique produces a very smooth surface.

Curved Tooth Files. A curved tool file is used widely in repair work to remove aluminum and solder from sheet metal and other smooth flat surfaces and on other soft metals such as brass and babbitt.

Machinist's Files. Machinist's files are widely used in machine shops in commercial and industrial facilities. (See Figure 1-23.) Most machinist's files are double cut and are used for both repair and production work. Table 1-2 lists common machinist's files and their uses.

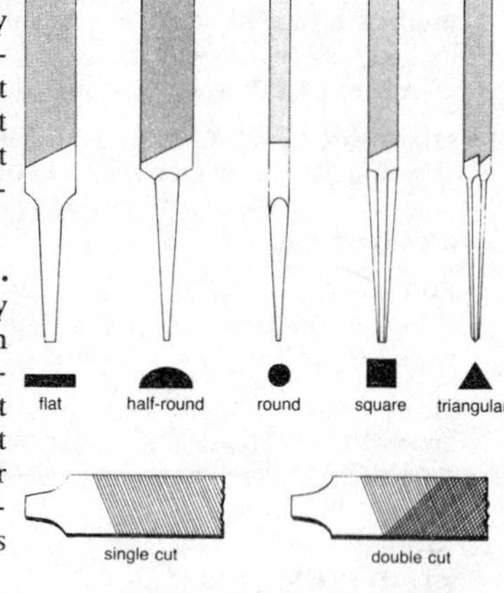

Figure 1-23. Common types of files.

Swiss Pattern Files. Considered fine finishing tools, Swiss pattern files are used for truing up grooves, keyways, notches, and other surface features. These files should only be used for delicate parts.

Wood Files and Rasps. These tools are used for dressing down and smoothing wood surfaces. Wood files are used for producing smooth finished work; rasps are coarser and are employed for rough work. Similar to other types of files, wood files and rasps come in various shapes and degrees of coarseness.

TABLE 1–2
Types of Machinist's Files

File Name	Comments
Mill file	Single-cut file used for draw filing to produce a fine smooth surface.
Flat file	Double-cut file that comes in a bastard cut for rough work and a second cut for finishing work.

TABLE 1–2 (*continued*)

File Name	Comments
Hand file	A thick rectangular file with parallel edges used to make slots and keyways.
Pillar file	Similar to the hand file, but with a narrow width and extra thickness, used for slot and keyway work.
Warding file	Thin rectangular file with a sharply tapered width of uniform thickness used for a variety of notching jobs.
Square file	With either a tapered or blunt blade, a file that is used for enlarging square holes or for slot and keyway work.
Round file	Tapered and used for filing curved surfaces.
3-Square file	A triangular cross-sectioned file used for internal filing and square corner clean-up work.
Half-Round file	File with one flat and one curved surface used for work on concave surfaces.
Knife file	A knifelike file—thick at one edge and thin at the other—used for clean filing of sharp cut-in angled surfaces or inside corners.

PLANES

A number of different planes are available for carpentry work, but all work on the same principle. With these tools, a knife projects from the bottom through an opening and shaves off material at various thicknesses.

Two planes most frequently used are the block plane and the jack plane. Other, more specialized, planes include rabbet, jointer, smooth, router, and bullnose planes. (See Figure 1-24.)

Block Plane. A low-angled plane, the block plane is usually operated with one hand. It is used primarily for finishing small surfaces and is desirable for fitting jobs.

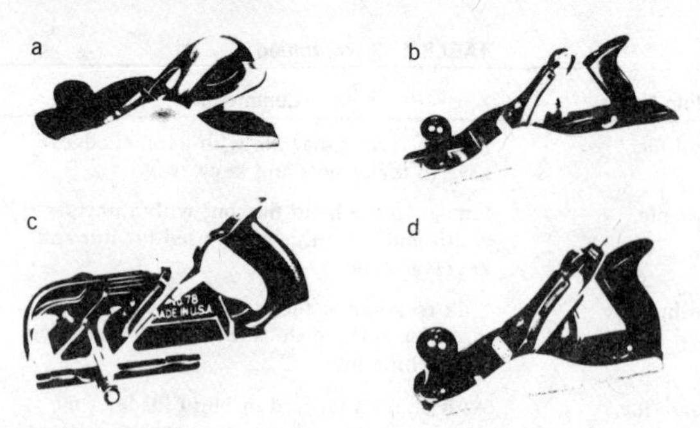

Figure 1–24. Planes: (a) block plane, (b) jack plane, (c) rabbet plane, (d) smooth plane.

Jack Plane. The jack plane is an all-purpose plane used for rough planing that gives lumber sufficient smoothness for further finishing. The most common size jack plane is a 14-inch length with a 2-inch blade. With this tool, one can smooth and join boards as well as perform other all-around work.

SAWING TOOLS

Saws are used for cutting various materials to size. The saws used most often in the maintenance mechanic trade are the backsaw, compass saw, coping saw, hacksaw, and handsaw. Most saws are available in various grades and weights, and selection and purchase should be made to meet individual needs. (See Figure 1-25.)

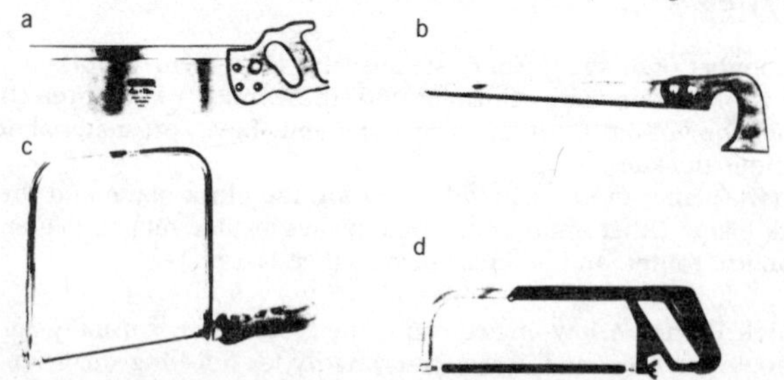

Figure 1–25. Saws: (a) backsaw, (b) compass saw, (c) coping saw, (d) hacksaw.

BACKSAW

A backsaw has a metal support strip along the tool's blade. Short backsaws are used for precision woodwork, longer saws in miter boxes. Backsaws range in size from 10 to 28 inches with 11 to 14 points to the inch—making for a fine finished cut.

COMPASS SAW

The primary use of a compass saw is to cut holes, such as those for electrical outlets. Compass saws can be purchased in lengths of 10, 12, and 14 inches (with 10 points to the inch). Most are equipped with a blade that can cut metal.

COPING SAW

A coping saw is a fine saw used for cutting coping joints and intricate patterns in wood or plastic. When you use a coping saw, remember that the teeth should always point away from the handle.

HACKSAW

The hand hacksaw is primarily used to cut metal and other hard material. Metal and plastic products, such as pipe, tubing, rod, plate, and forms, may be hand cut with this saw. The number of teeth on the saw blade determines the type of material that can be cut. (See Table 1-3.)

PROPER USE OF A HACKSAW

- When possible, clamp the workpiece in a vise.
- Have the broadest face of the workpiece facing up.
- Fasten the blade in the frame so that the teeth are pointing away from the handle.
- Hold the front of the frame, as well as the handle, when making your cuts.
- Make sure that all strokes are even and powerful. Do not twist the blade or lift it out of the cut.

TABLE 1–3
Blade Selection for Hacksaws

Number of Teeth per Inch	Diameter or Cross-Section of Material to be Cut	Material
14	1 inch or more	Aluminum, babbitt, bronze, brass, cast iron, cold-rolled steel, iron.
18	1/4 to 1 inch	Light and heavy angle iron, cast iron, drill rod, tool steel, general cutting.
24	1/16 to 1/4 inch	Brass pipe, BX heavy, heavy sheet metal, iron pipe.
32	under 1/16 inch	Sheet metal and tubing over 18 gauge, flush pipe, and BX light.

HANDSAW

The handsaw, perhaps one of the most easily recognized tools, is available with either a curved or straight back. Specifically designed for cutting wood, handsaws are divided into two groups: crosscut saw and ripsaw. The first is designed to cut wood across the grain, the second to cut wood along the grain.

As with other saws, the more teeth per inch, the finer the cut. Common crosscut saws have blades from 20 to 26 inches with 7 to 11 points to the inch. The most common ripsaw has a 26-inch blade with 5 1/2 or fewer points per inch.

PROPER USE OF A HANDSAW

- Keep your attention on the work and be sure to use properly sharpened saws.

- To assure accuracy and efficiency, use the right saw for the job.
- Make sure that the material being cut is well supported and free from nails or other obstructions.
- Start all cuts by drawing the saw backward using the thumb as a guide. Once a sufficient cut is made, use firm and powerful strokes.
- After use, wipe the blade of the saw with a thin film of oil to prevent rusting, and store the saw in an appropriate location.

SHEARING AND CUTTING TOOLS

Maintenance mechanics often need to cut sheet metal and other materials manually. This can be accomplished with hand snips or bench shears. Hand snips can be carried to the job site; bench shears require that the job be performed in the shop and are usually used only on large projects.

HAND SNIPS

There are a number of different types of hand snips, each designed for a specific type of cut.

AVIATION SNIPS

First used in the aviation industry for airframe work, aviation snips are used for cutting compound curves and other intricate shapes. There are three types of aviation snips: right, left, and universal. (See Figure 1-26.)

CIRCULAR CUTTING SNIPS

These hand snips are specifically designed for cutting intricate patterns and curves. One advantage of circular cutting snips is that they can make smooth cuts with no bending of the metal.

Figure 1-26. Aviation snips: right, left, and straight (universal).

DOUBLE CUTTING SNIPS

These snips are used to cut light gauge sheet metal piping.

HAWKBILL SNIPS

Hawkbill snips are designed to make curved cuts. As such, they have narrow curved blades.

STRAIGHT SNIPS

The straight snip, the most common sheet metal hand snip, is used for making straight cuts. Straight snips should be used for cutting sheet metal 22 gauge or thinner. (See Figure 1-27.)

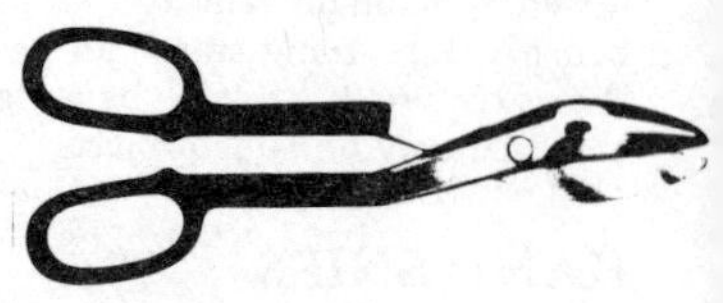

Figure 1-27. All-purpose straight snips.

HEAVY-DUTY SNIPS

The last hand snip to be considered is the heavy-duty snip, which is used for extra-heavy cutting. This snip is used for hand cutting sheet metal that is thicker than 22 gauge, and is capable of cutting both straight and curved lines.

PROPER USE OF HAND SNIPS

- Be sure the snips are sharp and well maintained. Blades should be taken apart and ground to an included angle of 85° and reassembled to a proper blade tension.

- Always select the proper snips for the job.
- Open the blades of the snips as far as possible and make long cuts.
- Cut to the outside of the layout line whenever possible.
- When cutting outside curves, begin by making a rough cut to within 1/8 inch of the line, and then finish the work in the manner described above.
- When making an inside curve cut, either drill or punch a hole in the center waste stock to permit the snips to start.

SCREWDRIVERS, PLIERS, PUNCHES, AND WRENCHES

This last group of tools is used for fastening and other operations.

SCREWDRIVERS

Screwdrivers are designed to work with slotted fasteners such as screws. Since there are a number of different types of fasteners, there are also a number of different types of screwdrivers. For large and heavy work, screwdriver bits are used in ratchet braces.

CONVENTIONAL SCREWDRIVERS

Conventional screwdrivers are used for driving slotted fasteners. They are available with shank lengths ranging from 1 1/4 inch to 12 inches. The blade also comes in various sizes so that it can be matched with the fastener. The tip should fit snugly into the head of the fastener and not be any wider than the diameter of the screw head. Otherwise, the surface around the fastener can be easily damaged. (See Figure 1-28.)

Figure 1-28. Standard tip screwdriver.

Conventional screwdrivers come in three basic sizes: small, medium, and large. Mechanics should have each size so that they

Figure 1-29. Magnetic tip screwdriver.

will be able to handle any size fastener. Some screwdrivers also come with a magnetic tip that allows for easier handling of the fastener. (See Figure 1-29.)

PHILLIPS SCREWDRIVERS

The Phillips screwdriver is designed for driving screws with a Phillips head. it is similar to the conventional screwdriver, except that the blade tip is cross-shaped (+). (See Figure 1-30.)

Figure 1-30. Phillips screwdriver.

SPIRAL RATCHET SCREWDRIVERS

Spiral ratchet screwdrivers are available in three different sizes and bits (slotted blade, Phillips, and countersink). Spiral ratchet screwdrivers are especially useful for driving a number of fasteners at one time. Unlike other screwdrivers that must be manually turned, the spiral ratchet design only requires the user to push on the handle. This action engages the spiral cut blade to turn within the mating housing. When pressure is released, an internal spring returns the blade to its original position. (See Figure 1-31.)

Figure 1-31. Spiral ratchet screwdriver.

PROPER USE OF SCREWDRIVERS

- Keep attention focused on the work.
- Only use screwdrivers that are in good condition and of the correct size and type.
- Locate the position of the hole and select the proper sized fastener.

- If there is no hole, bore a pilot hole slightly smaller than the diameter of the threaded part of the screw. For soft material (e.g., pine and basswood), the bore should be as deep as half the length of the threaded part of the screw. For harder material (e.g., oak, plastic, and hardboard), the hole must be almost as deep as the screw.
- Hold the screw with your free hand and begin driving it into the hole until the threads just grab hold of the stock.
- Drive the screw tightly in place.

SPECIALIZED SCREWDRIVERS

There are a number of special screw designs that will not accept a conventional or Phillips screwdriver, and specialized screwdrivers have been developed to fit these designs. Examples of these screw styles are hi-torque recess, torq-set recess, spline socket, slotted spanner, weld under, and weld over. Many of these special screw designs are found in sheet metal products, automobiles, electrical instruments, and specialty machines. (See Figure 1-32.)

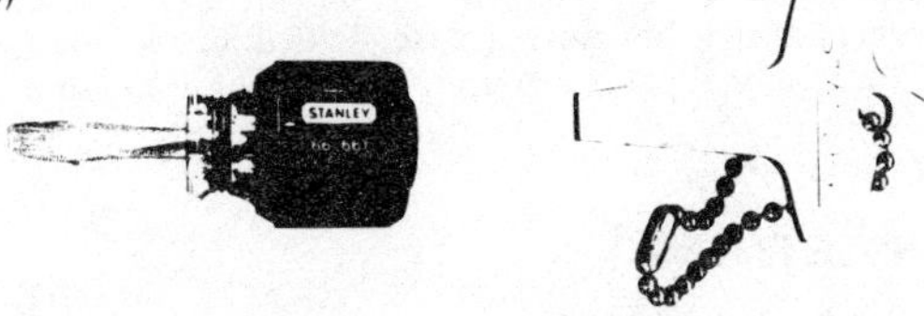

Figure 1-32. Stubby screwdriver and keychain screwdriver.

PLIERS

Pliers are used primarily for gripping. There are several types of pliers, each designed for specific uses. (See Figure 1-33.)

COMBINATION (SLIP JOINT) PLIERS

The most common type of pliers is the combination pliers, which is used primarily for gripping and bending wire and for removing wire or nails from stock material. Combination pliers come in lengths ranging from 8 to 10 inches.

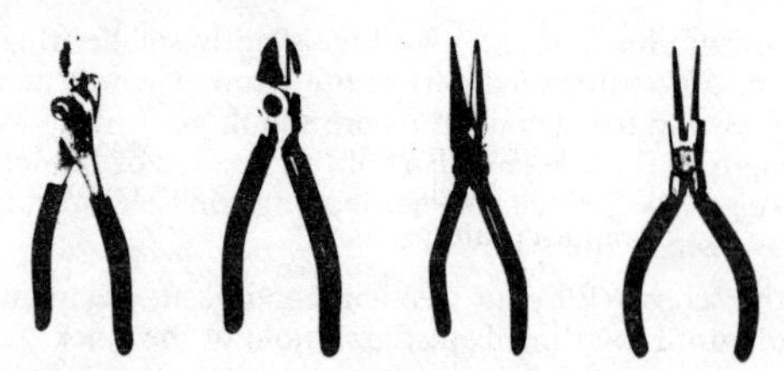

Figure 1-33. Pliers: slip-joint pliers, diagonal cutting pliers, long-nose pliers, and needle-nose pliers.

SIDE-CUTTING PLIERS

Side-cutting pliers are designed for the easy cutting of wires. At the working end of the tool there is a flat nose for gripping the wire, followed by a cutting bit.

DIAGONAL PLIERS

Another type of pliers commonly found in the mechanic's toolbox is the diagonal pliers. This pliers is similar to the side-cutting pliers, except that it has no flat nose—it is designed with a full cutting bit. Diagonal pliers are useful for cutting wire on a diagonal in tight places and can also be used for installing and removing cotter pins.

LONG-NOSE PLIERS

Long-nose pliers are used on electrical equipment. The long narrow nose of these pliers allows one to reach into tight places and makes it easier to loop and bend wire of all sizes.

NEEDLE-NOSE PLIERS

The jaws of needle-nose pliers are long and thin for holding small electronic parts or getting into very small places.

PUNCHES

In addition to punches used for layout work, such as center punches and awls, several punches are designed for nonlayout functions.

CENTER PUNCH

The primary use of center punches is for indenting or making small holes in metal surfaces. Center punches are quite useful in locating the position of holes to be drilled so that the center point of the drill can be started accurately.

NAILSET

These tools are used to "set" or drive the head of nails below the surface of the wood. The cavity formed is then filled and sanded smooth. Nailsets are available in tip sizes ranging from 1/32 to 1/8-inch diameter, with a length of about 3 3/8 inches. The size selected depends on the size of the nail used. (See Figure 1-34.)

Figure 1-34. Nailset.

SOLID PUNCH

The solid punch is used to punch holes in sheet material for the insertion of fasteners such as rivets. Solid punches have a solid or flat cutting end that comes in diameters ranging from 3/32 to 3/8 inch and are numbered from 6 to 10.

WRENCHES

Wrenches are used to tighten nuts and bolts and to set screws. They are usually classified according to the following categories: open-end and socket types, and adjustable and nonadjustable types. *Nonadjustable wrenches* are sized with fixed jaws; *adjustable wrenches* have one fixed and one sliding (adjustable) jaw. Adjustable wrenches come in small, medium, and large sizes. (See Figure 1-35.)

Figure 1-35. Adjustable wrench.

BOX WRENCH

The box wrench makes use of a socket to fit over bolt heads or nuts. The socket is round with twelve internal notches that hold onto the fastener and help prevent slippage. Box wrenches, often combined with open-end wrenches, are available in sizes ranging from 3/8 inch to 1 1/4 inch as well as in metric sizes. (See Figure 1-36.)

Figure 1-36. Wrenches: box wrenches, set of open wrenches, combination wrenches.

Box wrenches are often a favorite choice of mechanics because they make many jobs easier. They are used for a variety of jobs including work on trucks, heavy equipment, metal processing machinery, and various installation and servicing operations. A helpful feature of these wrenches is that they are made with one or both ends offset for access to difficult locations.

ALLEN WRENCH

The Allen wrench, also known as the *Allen hexkey,* is used for adjusting headless, hexagonal socket setscrews. (Setscrews, which are headless and slotted or with a hexagonal socket, are used to stop the movement of parts, such as a knob or the hub of a pulley on a shaft.) The body of an Allen wrench is L-shaped with a hexagonal cross-section to allow it to sit completely into the socket and not interfere with any other part. Allen wrenches come in sizes ranging from 0.028 inch to 1 inch and also metric sizes.

Allen wrenches are used for clamping and adjusting. For example, setscrews are often used to hold or clamp one machine element to another. On many machine tools, tooling and workpieces are moved into position for processing to within 0.001 or 0.0001 inch, and then locked into position with setscrews tightened by an Allen wrench. In addition, standard-sized Allen wrenches are also used for adjusting gas, diesel, and electric motor operation and speeds.

PIPE WRENCH

Pipe wrenches are used for turning pipes, fittings, and other round members. The jaws of a pipe wrench have sharp teeth that bite into the metal when pressure is applied to the handle. Therefore, this wrench should only be used when the surface has been protected with sheathing such as aluminum or copper sheet metal. (See Figure 1-37.)

Figure 1-37. Pipe wrench.

SPANNER WRENCH

The term "spanner" is used to describe a group of wrenches that span large distances. These distances often far exceed those accessible with standard box, adjustable, and socket wrench sets. A hook spanner wrench is used for ring nuts with square slots cut on the outside diameter. A pin spanner wrench is used on nuts having matching holes in their surface.

OTHER SPECIALIZED WRENCHES

T- and L-socket wrenches are recommended when extra force must be applied to a fastener. The T- and L-shaped handles allow the mechanic to apply additional torque for easier tightening or loosening.

2 Measuring and Testing Tools

Good maintenance service depends on procedures and requirements set forth according to machine design and specifications.

Predetermined and recognized standards and a variety of measuring, gaging, and testing devices aid the mechanic in achieving high quality work.

SPECIFICATIONS AND STANDARDIZATION

Specifications and standardizations used within the machine and maintenance trade include concepts such as size, location, operation, surface texture, material finishes, electrical requirements, and roundness. This section discusses standardization, dimensional and surface texture nomenclature, and quality control.

STANDARDIZATION

The need to interchange parts in machines and other equipment led to the standardization of parts and the establishment of the American National Standards Institute (ANSI). Virtually all engineering and technical societies responsible for standards and specifications are affiliated with ANSI, an umbrella organization that coordinates standards in a number of different technical fields. ANSI publishes and distributes standards and specifications pertaining to preferred practices, testing methods, safety, design, and sizes used by the vast majority of American industries. Table 2-1 lists the categories of ANSI standards.

The increase in international trade subsequently led to the establishment of international standards to meet the needs of the worldwide market. More than 40 countries have agreed to cooperate in the establishment of the International Organization for Standardization (ISO). Many of the ISO standards are the same as those provided by ANSI, though there are some that do not correspond to American standards. Any machine or maintenance mechanic involved with equipment and machinery that uses or produces products for international markets should check both ANSI and ISO standards.

TABLE 2–1
Categories of ANSI Standards

Code	Area
A	Construction
B	Mechanical
C	Electrical
D	Highway Traffic Safety
F	Food and Beverage
G	Ferrous Materials and Metallurgy
H	Nonferrous Materials and Metallurgy
K	Chemical
L	Textiles
M	Mining
MC	Measurement and Automatic Control
MD	Medical Devices
MH	Materials Handling
N	Nuclear
O	Wood
P	Pulp and Paper
PH	Photography
S	Acoustics, Vibration, Mechanical Shock, and Sound Recording
SE	Security Equipment
W	Welding
X	Information Systems
Y	Drawing, Symbols, and Abbreviations
Z	Miscellaneous

DIMENSIONAL NOMENCLATURE

It is important that the apprentice and journeyman understand the terminology used in specifying various sizes and fits. Those terms encountered in precision machine and tooling are discussed briefly here.

BASIC SIZE

The basic size of a part or product is the size from which the limits of size are determined by the use of allowances and tolerances. For example, the basic size of a machine part is the theoretical or nominal standard size from which variations are made. If the basic diameter of a shaft is 2 inches, this means that the theoretical size is 2 inches exactly, but in reality will vary within set limits—for example, +0.010 and −0.005 inch.

CLEARANCE

The clearance of a component is the difference in sizes between mating parts when the internal dimension of the female part is larger than the external dimension of the male part. Clearance is the opposite of interference.

INTERFERENCE

Interference refers to the difference in sizes between mating parts when the dimension of the internal female part is smaller than the dimension of the external male part. Interference is the opposite of clearance.

FITS

Fit describes the degree and quality of closeness with which the surfaces of parts are brought together. Because there are varying machine and product requirements, there are also different fit requirements. (See Table 2-2.)

Force Fits. Force, or shrink, fits make up a special type of interference fit that is characterized by a constant pressure on the mating parts.

Location Fits. Locational fits are those used for determining the location of mating parts only.

Running and Sliding Fits. Running and sliding fits are used to give the limits of clearance for running performance and take into account appropriate lubrication allowances.

TABLE 2–2
Classes of Fits

Fit	Classifications	Description
Force fits		
	FN1	Light drive fits for light assembly pressures.
	FN2	Medium drive fits for ordinary steel parts or for shrink fits on light sections.
	FN3	Heavy drive fits for heavy steel parts or shrink fits on medium sections.
	FN4 and FN5	Force fits for parts that are subjected to high stresses or where necessary heavy pressing forces are impractical.
Location fits		
	LC	Locational clearance fits for stationary parts that can be easily assembled or dissassembled.
	LN	Locational interference fits used for accurately locating parts and for components requiring rigidity and alignment.
	LT	Locational transition fits that fall between clearance and interference fits.

TABLE 2–2 (*continued*)

Fit	Classifications	Description
Running and sliding fits		
	RC1	Close sliding fits for accurate location of parts without perceptible play.
	RC2	Sliding fits for accurate location where parts may move and turn easily but are not intended to run freely.
	RC3	Precision running fits providing for the closest free running fits.
	RC4	Close running fits for running fits on accurate machinery having moderate surface speeds and journal pressures.
	RC5 and RC6	Medium running fits for higher running speeds and/or heavy journal pressures.
	RC7	Free running fits where accuracy is not critical and/or where temperature variations are found.
	RC8 and RC9	Loose running fits for wide commercial tolerances.

LIMITS

The dimensional limits of a part are the absolute minimum and maximum sizes the part can be.

TOLERANCE

Tolerance is the total amount of size variation that is allowed. It is the difference between the maximum and minimum limits of any dimension.

Tolerance is also defined as the amount that duplicate parts are allowed to differ in size to ensure needed accuracy without unnecessary refinement.

Bilateral Tolerance. Variation in size in both directions from the basic size is known as bilateral tolerance. It is expressed in both plus and minus dimensions, such as 2.50 ±0.001.

Unilateral Tolerance. The term unilateral tolerance is used to describe the total tolerance in one direction only—as either a plus or minus. An example of this would be 2.00 − 0.002 or 2.00 + 0.002.

SURFACE TEXTURE NOMENCLATURE

In cooperation with the American Society of Mechanical Engineers (ASME) and the Society of Automotive Engineers (SAE), the ANSI has developed specifications for surface textures. Surface textures, which pertain to smoothness and roughness, are important in a number of products, including a wide variety of machinery and machine parts. Technically, surface texture is defined as the repetitive or random deviations from the nominal surface that form a surface's pattern.

Listed here are the definitions of pertinent surface texture terms. Figure 2-1 illustrates surface characteristics and corresponding dimensional specifications.

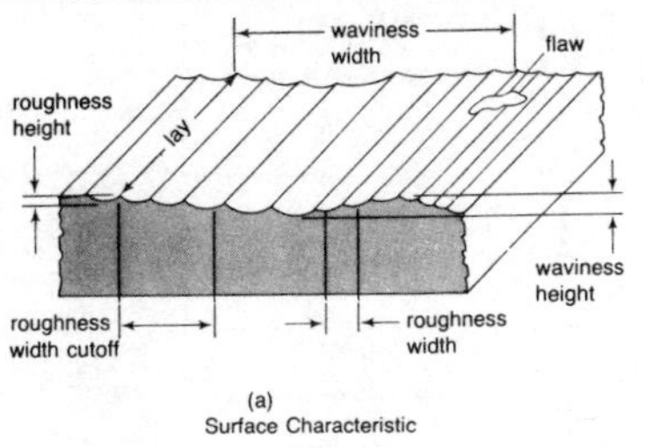

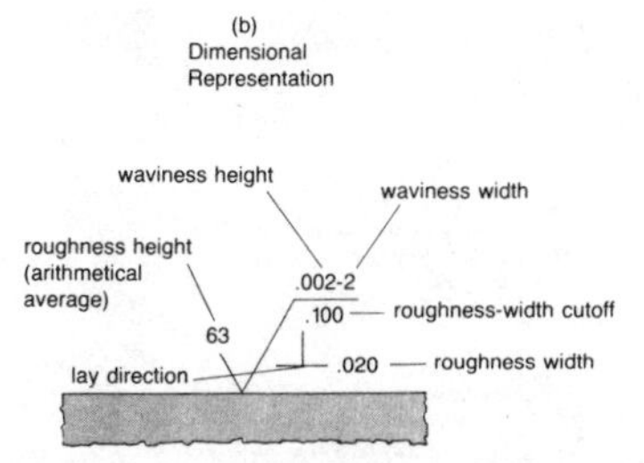

Figure 2-1. Surface characteristics and dimensional representation.

CENTER LINE

The center line is the reference about which roughness is measured. It is a line that is parallel to the direction of the profile within the limits of the roughness-width cutoff.

FLAWS

Any irregularities that are found on the surface at one location or at relatively infrequent or widely varying intervals are considered flaws. Examples include cracks, checks, scratches, and ridges.

Lay Symbol	Designation	Example
‖	lay parallel to the line representing the surface to which the symbol is applied.	direction of tool marks
⊥	lay perpendicular to the line representing the surface to which the symbol is applied.	direction of tool marks
X	lay angular in both directions to line representing the surface to which symbol is applied.	direction of tool marks
M	lay multidirectional.	
C	lay approximately circular relative to the center of the surface to which the symbol is applied.	
R	lay approximately radial-relative to the center of the surface to which the symbol is applied.	

Figure 2-2. Surface texture lay symbols.

LAY

Lay describes the direction of the primary surface pattern, which is determined by the production method. Figure 2-2 illustrates surface texture lay symbols.

NOMINAL SURFACE

The intended surface contour is known as the nominal surface.

PROFILE

The profile is the contour of a surface found in a plane that is perpendicular to the surface. Sometimes an angle other than 90° is used.

ROUGHNESS HEIGHT

Roughness height is a mathematically calculated standard expressed as the average deviation in microinches measured normal to the center line.

ROUGHNESS WIDTH

The roughness width is the distance parallel to the nominal surface between each succeeding peak or ridge forming the major pattern of the roughness. This feature is rated in inches.

ROUGHNESS-WIDTH CUTOFF

Rated in inches, the roughness-width cutoff is the greatest spacing that exists between repetitive irregularities that is included in the measurement of average roughness height.

SURFACE ROUGHNESS

The usual method of measuring surface roughness involves the calculation of the average deviation from the mean surface. In most cases, the total profile height of the surface roughness (peak to valley) will be equal to approximately four times the measured average surface roughness in microinches. Typical values for surface roughness produced by common processing methods is shown in Figure 2-3.

WAVINESS

Of wider spacing than the roughness-width cutoff, waviness is the widely spaced fluctuation of the surface texture. Waviness is caused by such factors as machine or work vibration, chatter, heat treatment, or a variety of warping stresses and strains.

WAVINESS HEIGHT

Waviness height is the distance between the wave peaks and wave valleys. The waviness height is taken normal to the nominal profile.

WAVINESS WIDTH

Measured in inches, the waviness width is the distance between successive wave peaks or successive wave valleys. When normally specified, the measurement given is the maximum allowed.

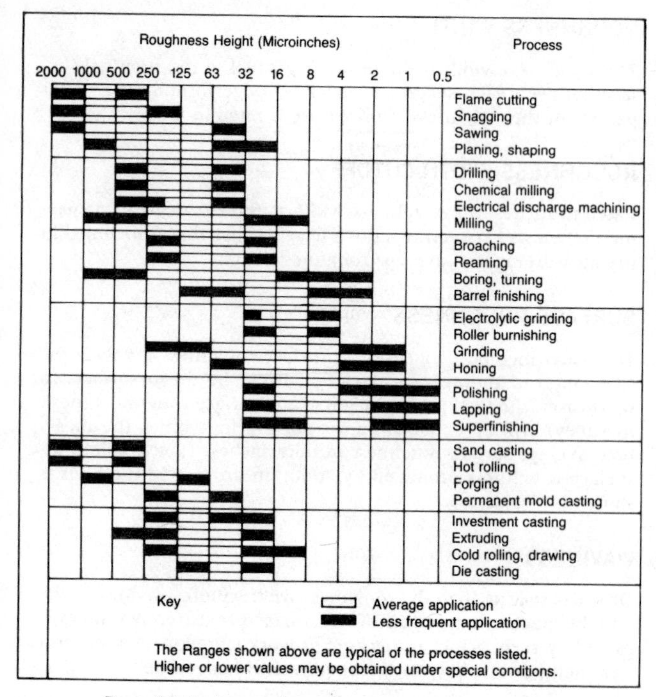

Figure 2-3. Surface roughness for common types of processes.

QUALITY CONTROL

The concept of quality control is usually associated with the production of manufactured parts. However, it is also applicable to maintenance and service activities. In order to maintain machines and facilities in high quality, inspections are performed using a variety of instruments, gages, and testing techniques. These tools and techniques are used to measure distance, angularity, surface texture, and roundness and to check allowances and detect imperfections in the structure of a part. The measured quality is then compared with a predetermined standard.

SPECIALIZED MEASURING INSTRUMENTS

Measuring instruments are used for a variety of service and repair jobs. Instruments used in the trade are classified as direct or indirect devices. *Direct measuring devices* are designed with a scale or some other means (for example, digital readouts) that allow the mechanic to read or determine the measure directly. Examples of direct measuring devices include scales, micrometers, and verniers. With *indirect measuring devices* it is not possible to obtain a direct reading; the device must be used in combination with some other device. Examples of indirect measuring devices are calipers, dividers, and telescoping gages.

ANGULAR MEASUREMENT

There are a number of instruments that can be used for measuring angles. Some, such as combination squares, toolmaker's telescope, and comparators, are used for both angular and linear measurement, but others, such as the protractor, sine bar and plate, and gages, are specifically used for angular measurement. The common unit used to express angularity is the degree, which represents 1/360 of a full rotation.

GAGES

A gage is any instrument against which measurement is compared to a given standard or system. There are two basic types of gages used to make angular measures.

Angle Gage Block. Angle gage blocks are built up of blocks wrung or slid together to produce angles of different sizes. These blocks are available in various accuracy categories and can measure angles to within 1 second (0° 0′ 1″), or 1/360 of a degree.

Angle Gage. The angle gage consists of a series of leaf gages of differing angles that can be inserted or placed parallel to the angle in question. This type of gage is commonly used for inspection jobs that involve measuring angles between 1° and 45°. (See Figure 2-4.)

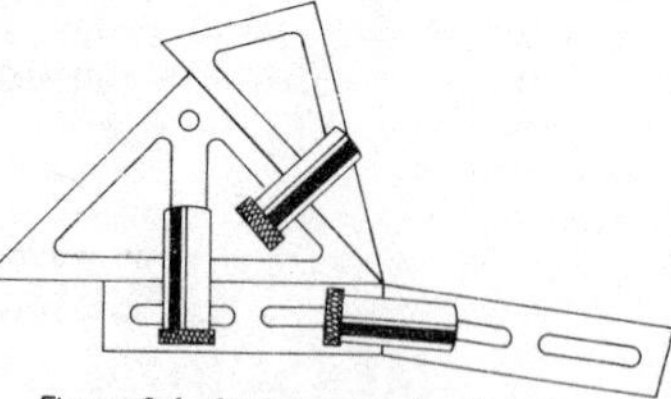

Figure 2-4. Angle gage being held in place by gage pins.

PROTRACTORS

Bevel Protractor. The bevel protractor consists of two adjustable blades that conform to angular surfaces. Position the device and adjust the blades and then compare the angle to a graduated protractor or angle gage.

Universal Bevel Protractor. The universal bevel protractor is usually equipped with a vernier angular scale. (See Figure 2-5.) This instrument is capable of making measures to the nearest half degree (1/2° or 30′).

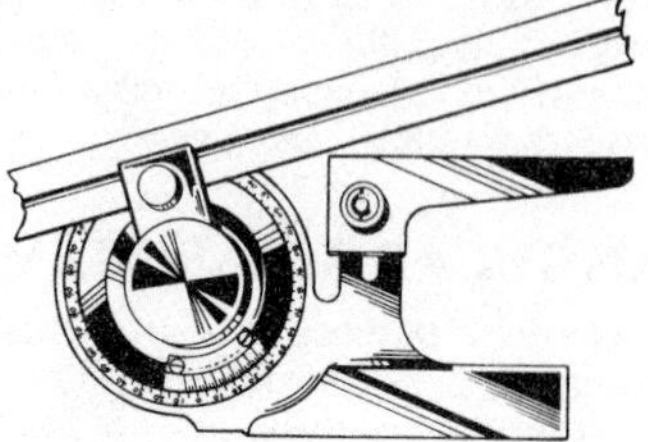

Figure 2-5. Universal bevel protractor.

SINE BAR AND PLATE

The sine bar is an accurate angle-measuring instrument used in the inspection of tools, jigs, dies, tapers, and fixtures. This instrument is made of a highly accurate ground steel bar and two pins of like diameters. The centers of the pins are located at precisely predetermined lengths ranging from 5 to 20 inches (127

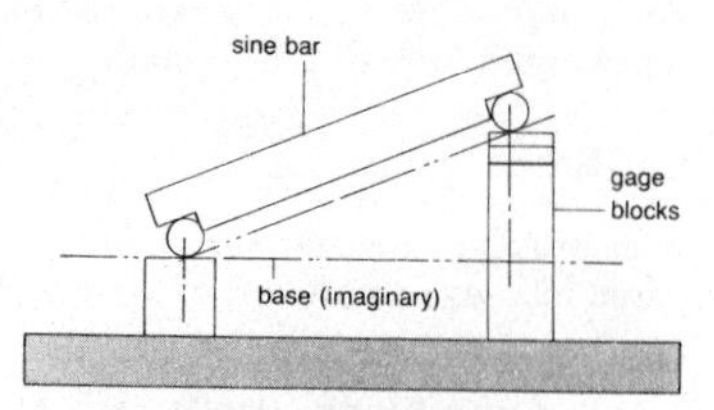

Figure 2-6. Use of a sine bar in combination with gage blocks.

to 508 mm) at 5-inch intervals. The use of sine bars is based upon trigonometric functions of sine and/or the relationship of sides and angles in a right triangle. (See Figure 2-6.)

Sine plates are highly accurate and finished cylindrical pins that are attached to a precision surface plate. Angular measures made with gage blocks are employed with the sine bar. Often, magnetic sine plates are used to hold a part for machining or inspection.

ELECTRICAL MEASUREMENT

Mechanics use electrical measuring instruments to measure quantities such as line voltage and power consumption and also to measure temperature by thermocouple voltage.

GALVANOMETER

The most common instrument used to measure direct current (DC) generating devices such as those found in batteries and many transformers is the galvanometer. (See Figure 2-7). Readings are taken through input terminals that send the current through a coil around a magnet. As the current flows, it produces a magnetic moment and tends to rotate the coil, which, in turn, reacts with a hairspring. The hairspring is attached to a pointer along a scale, so that a direct reading can be made.

In addition to measuring current, the galvanometer is converted to a DC voltmeter, ammeter, or ohm-

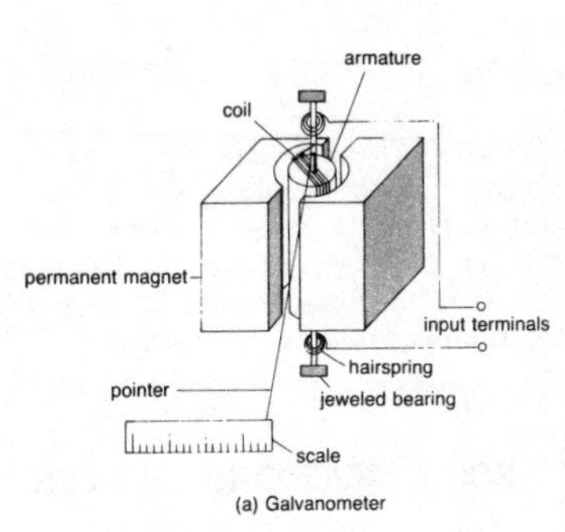

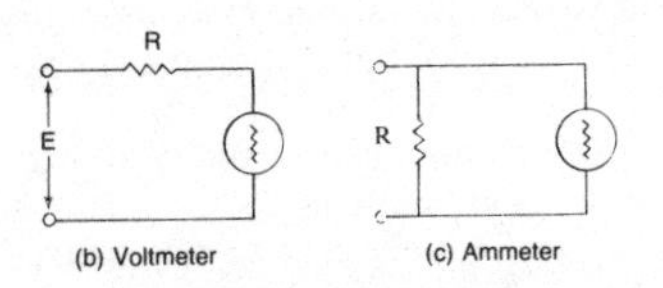

Figure 2-7. Electrical measuring instruments.

meter by using Ohm's law: $IR = E$, where I = current, R = resistance, and E = electrical potential (voltage).

OTHER DC MEASURING DEVICES

Ammeters are used to measure electrical resistance, and are made by placing the resistance in parallel with the galvanometer. The typical ohmmeter consists of a battery, adjustable rheostats, and an ammeter in series with the resistance being measured. An electric voltmeter, known as a VTVM (vacuum tube voltmeter), traditionally has been used to measure circuit current, resistance, and voltage, and is still in common use today.

AC MEASURING DEVICES

Alternating current (AC) measurement is important when maintaining and servicing electric generators and motors. VTVMs can be modified to take measurements in AC circuits. Other devices—*electrodynamic meters*—though not as sensitive as the VTVM, are used to measure alternating current in some shops. Power is expressed in terms of wattage and measured by *wattmeters*.

OTHER ELECTRICAL/ELECTRONIC MEASURING DEVICES

Advances within the microelectronics industry have made it possible to provide small and accurate electronic testing equipment. One such device is the *pocket multimeter*, which usually has a liquid crystal display (LCD). Typical pocket digital multimeters are used to test electronics and electrical appliances, machines, instruments, and controls. They can measure both AC and DC voltage as well as resistance.

Frequency, phase, and time measurements are usually accomplished with the aid of a cathode ray oscilloscope (CRO). This is perhaps one of the most versatile and widely used testing instruments used in the laboratory. However, most maintenance activities will not require this level of sophistication.

FLUID AND FLUID FLOW RATE MEASUREMENT

Fluid level measuring instruments are used to determine and control the amount of liquid in storage tanks and other vessels. In production facilities, they are also used to indicate the location of two interfacing liquids having different properties. Large storage tanks often make use of a calibrated tape or chain that is attached to a float, which floats, or rides, on the liquid's surface. Smaller vessels often employ a *fixed displacer*.

Fluid flow rate is usually expressed in terms of volumetric units per time. For example, the flow rate of gases is expressed in cubic feet (ft^3) or cubic meters (m^3) per minute (min), written as ft^3/min or m^3/min. Steam flow is expressed as pounds (lb) or kilograms (kg) per hour (h), written as lb/h or kg/h. Liquid flow is expressed as gallons (gal) or liters (l) per minute (min), written as gal/min or l/min.

PITOT TUBE

The primary use of the pitot tube is to determine fluid velocity. It can be used to measure incompressible or low speed flow as well as compressible flow. A pitot tube is made up of two concentric tubes that lie across some type of differential pressure-measuring device. It measures the stagnation pressure—that is, the impact or total pressure—of a flowing fluid, and is therefore sometimes called an *impact tube*. When total pressure is to be measured, a pitot tube system will consist of a primary sensing element mounted on a support, pressure connecting lines, and a pressure indicating device. The connecting lines and indicating device are secondary elements and are not usually considered part of the pitot tube.

VENTURI TUBE

Average flow rate is usually measured by a venturi tube, a device based on the differences in pressure as fluid passes through a narrowed throat. Basically, a venturi tube is a short straight pipe section, known as a "throat", that lies between two tapered sections. Local pressure varies in the area of the constriction. When the throat is attached to a manometer (or recording instrument), the drop in pressure as the fluid flows through the wider area is measured and the flow rate calculated. The major advantage of using a venturi tube is that no more than 10 to 20 percent of the difference in pressure between the inlet and the throat is permanently lost.

METERING PUMPS

The metering pump is an accurate means of measuring and controlling fluid flow rates. Metering pumps are designed with either a propeller or turbine situated in the line and are mechanically geared to a counter. As the fluid passes through the line, the propeller rotates and the flow rate is counted.

OTHER FLOW MEASURING DEVICES

Airflow readings are easily made with the *hot wire anemometer*. Two meters that are gaining wider acceptance are the *electromagnetic flowmeter* and the *vortex shedding meter*. The first instrument does not have to be inserted in the flow stream, while the second acts as an obstruction in the line itself.

FORCE AND TORQUE MEASUREMENT

Force is measured by the deflection of an elastic element; torque is a product of a force and the perpendicular distances to the axis of rotation. Most accurate force and torque measuring instruments make use of a *strain gage*, in which electrical resistance changes along with applied strain. The relationship of gage-resistance change to input variation is analyzed and calibrated to force and torque measures.

TORQUE WRENCH

The torque wrench allows the mechanic to tighten fasteners to within given torque allowances by reading a torque gage or meter on the wrench. This device is often used on machinery and equipment such as vehicles, cranes, and cutting centers.

LINEAR MEASURING INSTRUMENTS

The basic instrument used for linear (displacement) measures is the steel rule. Rules can be purchased in lengths ranging from 4 inches to over 100 inches, with the most common being 6 and 12 inches. Usual inch graduations are 8ths, 16ths, 32nds, and 64ths. Other rules are available in 10ths, 50ths, and 100ths of an inch. Rules can be found with customary and/or metric scales with metric scale divisions expressed in millimeters (mm).

COMBINATION SQUARE

Already discussed in the previous chapter as a layout tool, the combination square consists of a rule with a center head, a square head, and a protractor attachment. The square head is used for measuring hole depths and checking squareness and 45° angles as well as for linear measures. The protractor is used for angular measures, and the center head is used to find the center of cylindrical objects.

MICROMETER

The micrometer is based on the principle of the calibrated screw. If the micrometer screw is turned one revolution, it moves 0.025 inch. In turn, one revolution of the thread is divided into 25 equal parts which represents 1/25 of 0.025 inch, or 0.001 inch.

In reading micrometers, it is important to count the number of revolutions made by the screw and add to it any fraction of a revolution. To aid in this process, micrometers are equipped with a scale on the edge of the thimble and barrel. For example, in Figure 2-8, the zero on the thimble does not exactly coincide with the reference line of the barrel, which notes that the thimble has not traveled one complete revolution beyond the preceding graduation on the barrel. To make the micrometer reading, the thimble units in thousandths are added to the barrel units in 0.025:

0.250 (barrel) + 0.021 (thimble) = 0.0271 inch

All micrometer calipers are the same in principle, though they may vary in size, frame construction, type of gaging anvils, and unit of measure (U.S. Customary or Metric).

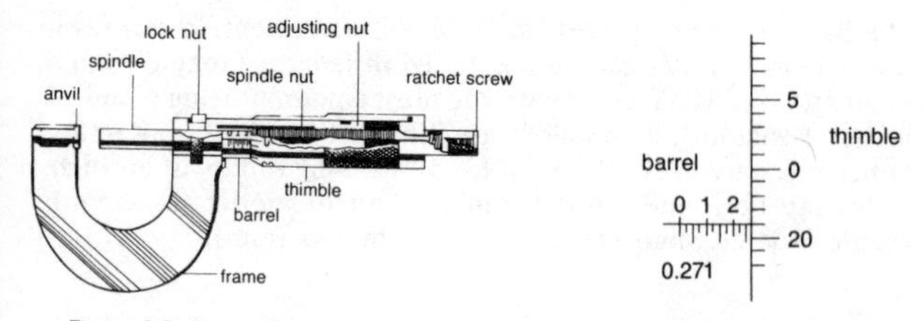

Figure 2-8. Parts of a micrometer caliper and detail showing a micrometer reading of 0.271 inch.

Outside Micrometer Calipers—"Mike". Perhaps the most extensively used micrometer is the outside micrometer calipers. Usually referred to as the "mike," it is popular because of its convenient size, accuracy, ease of operation and reading, range of measurement, and economical cost.

1. *Ball-Anvil Micrometer Caliper.* A special outside micrometer designed with a ball-shaped anvil, the ball-anvil micrometer is used for inspecting accurately the diameter of cylindrical parts such as tubing.
2. *Sheet-Metal Micrometer Caliper.* This is a structurally rugged micrometer that has been specifically designed for gaging sheet metal in rolling mills. Features include a heavy duty frame, large coarse pitch measuring screw, wear-resistant anvils, and corrosion-resistant parts.
3. *Bench Micrometer.* The bench micrometer is designed with finer threads and is capable of directly measuring to within the nearest ten-thousandths of an inch. It is used with wire and other products demanding exceptional accuracy.

Inside Micrometer. The inside micrometer is capable of inspecting large bores as great as 100 inches in diameter. When it is necessary to measure larger holes, extension rods are used. In addition to measuring the diameter of bores and recessed areas, the inside micrometer can also be used as a height gage, though this is seldom done.

Inside Micrometer Calipers. A variation of the outside micrometer, the inside micrometer caliper is used to inspect small inside dimensions. This measuring instrument has two jaws with gaging nibs that are hardened and ground to a given diameter. They are designed with a small locking nut that allows one to hold the dimension until the reading is noted. Unlike other micrometers, the inside micrometer caliper is limited to dimensions ranging from 1/2 inch to 2 inches.

Micrometer Depth Gage. The micrometer depth gage is a precision instrument used for inspecting holes, slots, and recessed areas. The shoulder of the instrument is held perpendicular to the centerline of the hole as the rod is adjusted for measuring. The screw in these depth gages has a 1/2 inch travel. Measures requiring greater depth will use extension rods that are graduated at 1/2 inch intervals.

CARE AND ADJUSTMENT OF MICROMETERS

Micrometers are precision instruments, and a machinist or maintenance mechanic should be able to properly care and service these tools to ensure their working accuracy.

- Never use the micrometer as a snap gage, for there is a good possibility that the springing of the tool will make it unsuitable for further use.
- Do not store micrometers with the anvils closed; always keep them slightly separated. One reason for this is to prevent corrosion, which can occur when any moisture is present and metals are in contact.
- Use only a drop of oil to lubricate the micrometer screw.
- Use micrometers only on stationary parts. For a reading on a part that moves, be sure that the machine is turned off and all motion has stopped.
- Clean the micrometer screw frequently with benzene or some similar agent.

- Check the spindle and anvil for wear and be sure they are flat and parallel to within 0.0002 inch.
- Never tighten the anvil closed during inspection or storage. When taking a reading, the anvil should close until it just comes in contact with the surface.

VERNIER INSTRUMENTS

Measuring instruments that make use of the vernier scale are based on the relationship between two different graduated scales. Vernier instruments are found in a variety of forms, including micrometers, calipers, height gages, and depth gages. The major advantages of these instruments are their accuracy, their adaptability to different types of work, and their capability of a wide range of measurement.

To read a vernier scale, such as that on a vernier micrometer, determine which vernier graduation on the barrel coincides with a graduation on the thimble and then add the amount in ten-thousandths (0.0001) to the regular micrometer reading. For example, in Figure 2-9, the micrometer readings are 0.2950 and 0.449 inch.

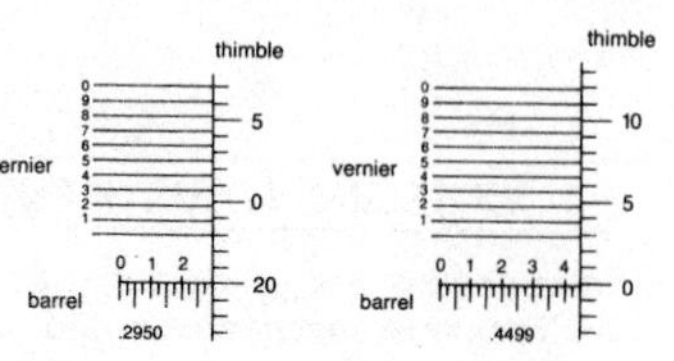

Figure 2-9. Vernier micrometer readings.

PRESSURE AND VACUUM MEASUREMENT

Pressure is defined as force per unit area exerted by a fluid (gas or liquid). It is measured in relationship to atmospheric pressure, which is 14.7 pounds per square inch, or 14.7 lb/in^2. One of the most familiar instruments used to measure atmospheric pressure is the mercury barometer. However, the most common fluid pressure device is the *Bourdon-tube gage* (See Figure 2-10.).

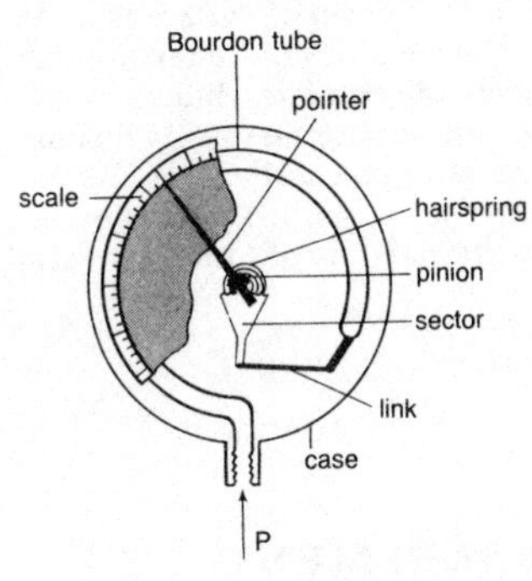

Figure 2-10. Bourdon-tube gage.

The Bourdon-tube gage consists of a flattened tube made from a spring metal (usually bronze or steel) that is bent into a circular form. As pressure increases inside the tube, it tends to unwind the spring. The spring is connected to a point via a pinion. Thus, as pressure increases and decreases, the spring tightens and unwinds, moving the needle in a clockwise or counterclockwise direction.

Other types of gages used to measure fluid pressure are the *bellows* and *diaphragm gage*. Both make use of a diaphragm or elastic member that deflects as pressure changes. The diaphragm is usually made out of brass, stainless steel, leather, neoprene, or rubber.

ROUNDNESS MEASUREMENT

Most roundness dimensions can be checked by a diameter or radius measure. Instruments such as micrometers, vernier calipers, and gages are helpful in checking roundness. However, when great accuracy is needed, such instruments are insufficient. *Profiling instruments* use a tracer device and recording mechanism to measure roundness accurately. Any change or variation in stylus movement produces an electrical signal. With this instrument roundness measures can be checked to within the nearest 0.000003 inch (0.000076 mm).

SURFACE TEXTURE ASSESSMENT

The most frequent and long-standing method to assess surface texture is visual inspection. As a general means of evaluation, this technique may be sufficient. However, there may be times when a more accurate method is needed.

Calibrated blocks or gages are used to evaluate surface texture. The blocks are made with varying degrees of surface texture that can be visually compared and come in sets of varying types and degrees of surface texture. Of all the surface texture evaluation tools available, this is perhaps the most practical and most commonly used by mechanics. Other, more sophisticated, instruments available include optical flats, profilometers, and autocollimators.

TEMPERATURE MEASUREMENT

Within the industrial and commercial setting, temperature is expressed in two common scales: Fahrenheit and Celsius. The Kelvin scale is sometimes used, but primarily in laboratory situations. Both the Fahrenheit and Celsius systems are based upon the boiling and freezing points of water.

Temperature can be measured in several different ways, some more useful than others. There are seven different principles employed for temperature measurement, and each involves the use of a specific type of measuring device. (See Table 2-3.)

The most common temperature measuring device used in the maintenance field is the *mercury-in-glass thermometer,* which has a useful temperature range of −30 to 900°F (−35 to 500°C). For industrial situations, the thermometer will usually be encased in a protective well and case. Many commercial thermometers come with a threaded union fitting that can be installed in a line or vessel while under pressure for determining line temperatures.

The majority of *thermostats* used for building climate control systems are bimetallic gages. When temperature readings must be transmitted over distances up to 1000 feet, a *pneumatic transmitter* is recommended. High temperatures, such as those used in industrial heat treating processes, are accurately measured with *thermocouples; resistance thermometers* are particularly useful for measuring low temperatures. *Optical radiation pyrometers* are normally applied to temperatures over 1000°F (540°C).

TABLE 2–3
Temperature Measurement

Principle	Instrument	Description
Thermal expansion of a gas	Gas thermometer	At a constant volume, the pressure of a gas is directly proportional to its absolute temperature.
Thermal expansion of a solid or liquid	Mercury thermometer, bimetallic element	All substances tend to expand with an increase in temperature. A change in temperature causes a change in length or volume.
Vapor pressure of a liquid	Vapor-bulb thermometer	Temperature increases bring about an increase in vapor pressure of all liquids.
Thermoelectric potential	Thermocouple	When two dissimilar metals are in contact, voltage that is dependent on the temperature at the junction is generated.
Variation of electrical resistance	Thermistor, resistance thermometer	The resistance of electrical conductors changes with temperature.
Change in radiation	Optical pyrometer, radiation pyrometer	Matter radiates energy proportional to the fourth power of its absolute temperature.
Change in physical or chemical state	Seger cones, tempilsticks	Temperatures at which substances melt or undergo chemical reaction is known and is reproducible.

VELOCITY AND ACCELERATION MEASUREMENT

Velocity is the time rate of displacement, and is a critical measure in many pieces of machinery and operating equipment. One of the instruments most widely used to measure velocity is the *tachometer*, which provides for a direct measure of angular velocity. This instrument is coupled to the rotating shaft or component being measured. Voltage is generated in an armature coil and is directly proportional to the speed. The voltage is then converted into either an analog or digital scale readout.

Vibration velocity is measured by a coil that is moved relative to a magnet. Known as vibration *acceleration pickups*, they operate on the same principle as a strain gage or piezoelectric element. They are critical instruments in determining excessive vibration and chattering in many production machines and equipment.

3

POWER TOOLS

Power tools are designed to accomplish the same jobs as hand tools, but with greater ease and speed. However, hand tools should not always be relegated to second best, for there are still many tasks that can be accomplished more easily with hand tools.

There are two major categories of power tools: stationary and portable. Stationary power tools include all electric, hydraulic, and pneumatic tools that, once positioned, will not be moved, or in other words, cannot be taken to the work site. Portable tools, in contrast, include everything from small electric hand tools to heavy-duty equipment that is portable and can be taken to the work site. This chapter discusses common power tools used within the machine and maintenance mechanic trade.

STATIONARY POWER TOOLS

Mechanics that work in facilities with a shop area often have stationary power tools available to them. Industrial maintenance shops are virtually quite sophisticated and have a wide variety of machinery and tooling. Smaller shops found in commercial and business buildings require less specialized and a smaller number of machines.

Most stationary power tools are used for drilling, grinding, milling, sanding, sawing, or turning operations.

THE DRILL PRESS

The drilling machine is commonly referred to as a drill press. It is used to produce, enlarge, or finish holes by various machining operations, including drilling, reaming, boring, counterboring, countersinking, and tapping. Other operations performed by a drill press are chamfering, spotfacing, and counter drilling. (See Figure 3-1.)

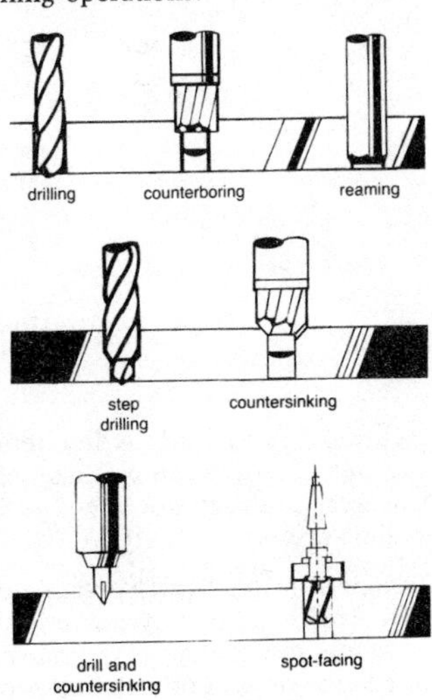

Figure 3-1. Common drilling operations.

TYPES OF CUTTING TOOLS ON DRILLING PRESSES

Twist Drill. The basic cutting tool used on drill presses is the twist drill. Twist drills are designed to resist breakage and damage caused by high torque and impact forces. One of the most important factors in twist drill design is the point angle. Generally, the harder the material to be cut, the larger the point angle should be. Recommended point angles are listed in Table 3-1.

TABLE 3–1
Point Angles for Drilling Different Materials

Point Angles (degrees)	Material
60	Aluminum, die castings, plastics, wood, and other soft material
60–118	Magnesium alloys
90–110	Cast iron, malleable iron, aluminum, plastics, slate, marble, wood, and hard rubber
100–118	Copper and brass
118	General purpose materials, carbon steels, annealed steels, cast iron, malleable iron, and medium-hard pearlite
125	Stainless steel, alloy steels, drop forgings, hardened steels, and materials that are difficult to cut
150	Tough alloy steels, 7–13% manganese steel, and other very hard materials

Reamer. The reamer is frequently used in drilling operations. Reamers are precision cutting tools used to machine smooth finish holes to exacting dimensions. Reaming is done after rough drilling or boring.

Other Cutting Tools. Sometimes existing holes must be modified. This is frequently accomplished with cutting tools such as counterbores, countersinks, and spotfacers. Counterbores are used to enlarge holes. Countersinks produce a cone-shaped opening at the surface of the workpiece. Spotfacers machine a smooth flat surface that is 90° to the hole's axis; this is often necessary when bolts, nuts, cap screws, and other types of mechanical fasteners must sit flat on the surface.

TYPES OF DRILLING MACHINES

There is a wide variety of drilling machines available. Some, including radial, turret, and multiple spindle drilling machines, are specifically designed for production purposes; others are more suited for jobs performed by the machine and maintenance mechanic. Some drill presses are small and require only a fraction of a horsepower to function; others are heavy-duty presses that make use of more than 50 horsepower. The two most common types of drilling machines found in the machine and maintenance mechanic shop are the upright drilling machine and the sensitive drilling machine.

Upright Drilling Machine. One of the most common and recommended types of drilling machines found in the shop setting is the upright drilling machine. Considered an all-around, general purpose machine, it is made with a vertical column that can be round or box-shaped. A sliding table is attached to the column. Tables on box-shaped columns move only in a vertical direction; those attached to round columns not only move vertically but can also rotate about the column.

The tables are either circular or square with machined holes to accommodate clamps or bolts. The table's ability to move and serve as a base for workpiece fastening enables the drilling of large and small parts as well as workpieces that have varying shapes. The drills and chucks used are held in the drill spindle by an American Morse Taper. Once the tool is in place, the spindle housing can be lowered or raised for machining.

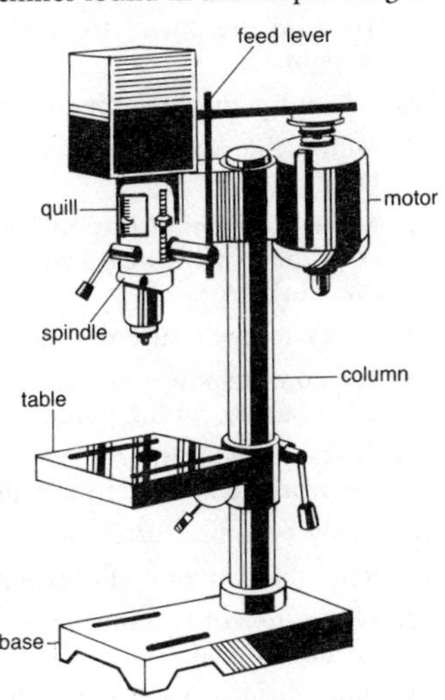

Figure 3-2. Sensitive drilling machine.

Sensitive Drilling Machine. This type of drilling machine is used for drilling small holes and is limited almost exclusively to the shop setting. The sensitive drilling machine is similar in appearance and operation to the upright drilling machine, except that it does not come equipped with an automatic power feeding machine. Most bench drill presses used today are of this type. (See Figure 3-2.)

Use of a Drill Press

1. Select the proper cutting tool (e.g., drill or reamer).
2. If a straight shank tool is used, mount it in a drill chuck. If a tapered shank is used, insert it directly into the spindle or in a sleeve and then into the spindle.
3. Turn on the power to make sure that the tool is running straight.
4. Adjust the machine for correct speed.
5. Mount the workpiece in a vice, clamp, or fixture on the table.
6. Move the point of the cutting tool with the handle feed lever so that it will line up with the appropriate layout mark on the workpiece.
7. Check to make sure that the depth of cut will be correct.
8. Turn on the power and feed the tool into the stock. If appropriate, apply cutting fluid and begin machining.
9. Check to see if the cut is true with the layout. If not, move the workpiece so that the error is corrected.
10. Apply even pressure.
11. Raise the cutting tool occasionally to clear the hole of chips.
12. When the cut is complete, raise the tool from the work and turn off the power.
13. When drilling large holes, drill a pilot or lead hole at a diameter equal to or slightly larger than the web of the drill.

GRINDING MACHINES

Grinding consists of the abrading, or wearing away, of material by an abrasive material.

GRINDING WHEELS

The cutting edge of a grinding wheel consists of a large number of abrasive particles or grains. The abrasive grains are held in place by various bonding agents and are press-formed in various disc designs.

Abrasives. The two most common materials that grinding wheels are made of are aluminum oxide (Al_2O_3) and silicon carbide (SiC). Both are manufactured materials that are hard, tough, and wear-resistant. Natural abrasives, such as sandstone, corundum, and emery, are seldom used because of their nonuniform properties.

The most widely used abrasive material is aluminum oxide, which is gray to gray-purple. Wheels made from this abrasive are best for grinding plain and alloy steels. Though silicon carbide is a much harder material, it is less often used because of its significant brittleness. However, silicon carbide wheels are preferred for grinding nonmetalic materials such as carbides, stones, and porcelain enamel.

The proper selection of abrasive grain or grit size is determined by the type and rate of material removal and the surface finished desired. Coarse grains should be used for grinding ductile metals, where excess material must be removed rapidly. Fine grain abrasives are best suited for the removal of hard and brittle materials, where the removal rate is low but the finish is fine. Common abrasive grit sizes are presented in Table 3-2.

Bonding Agents. Grinding wheels are bonded by either organic or inorganic agents. Organic bonds include materials such as rubber, shellac, and resinoid; inorganic bonds include glass, clay, oxychloride, and sodium silicate.

The most common type of grinding wheel found in the shop setting is the vitrified wheel. Vitrified bonding is an inorganic bonding process that uses clay, glass, porcelain, feldspar, or other similar ceramic materials. These wheels are brittle and rigid, but resistant to water, oil, and acids. Organically bonded wheels are used only in situations in which the wheels are exposed to extreme bending forces, and in these cases some reinforcement, such as a fabric or filament, is usually incorporated.

TABLE 3–2
Common Abrasive Grit Sizes

Category	Grit Sizes
Extra Fine	600
	500
	400
	320
	280
Fine	240
	220
	180
	150
	120
Medium	100
	90
	80
	70
	60
	54
	46
	30
Coarse	24
	20
	16
	14
Extra Coarse	12
	10
	8
	6

Coding of Wheels. Grinding wheels are identified by a characteristic code, a uniform marking system accepted throughout the industry. The performance of individual grinding wheels varies from one manufacturer to the next but should be uniform within a single manufacturer. (See Table 3-3.) For example, a wheel designated as 48-C-90-Q-12-V-17 tells us that the manufacturer's abrasive symbol is *48*; the abrasive type is *silicon carbide,* which has a fine grit size of *90*; its grade is a hard *Q*; the structure is *12*; its bond type is *vitrified*; and the *17* is optional and may be used for the manufacturer's private factory records.

TABLE 3–3
Uniform Marking System for Grinding Wheels

Marking Categories (in order)	Codings
Prefix (optional)	Manufacturer's symbol for abrasive type
Abrasive material	A: aluminum oxide C: silicon carbide
Grit size	Coarse: 10 through 24 Medium: 36 through 60 Fine: 70 through 180 Very Fine: 220 through 600
Grade	Soft: A through H Medium: I through P Hard: Q through Z
Structure (optional)	Dense to Open (1, 2, 3, 4, 5, 6, 7, 8, 9, 10, 11, 12, 13, 14, 15, etc.)
Bond type	V: vitrified S: silicate B: resinoid R: rubber E: shellac O: oxychloride F: reinforced
Manufacturer's record (optional)	Manufacturer's mark to identify the wheel

In addition to the uniform marking system used for wheels, there are nine standard grinding wheel shapes. Grinding wheels that are straight or recessed on one or both sides are used for cylindrical, surface, off-hand, and snag grinding. Tapered wheels are usually reserved for snag grinding. The saucer, or saw gummer, wheel is used for sharpening saws; the dish-shaped wheel for grinding in narrow places.

TYPES OF GRINDING MACHINES

The types of grinding machines that are of importance to the machine and maintenance mechanic can be broken down into two major groups: general purpose tool grinders and precision grinding machines.

General Purpose Tool Grinders. One of the most frequently used grinders found in the maintenance shop is the tool grinder, used mainly to maintain sharp cutting tools. Tool grinders may be bench-top or pedestal mounted. There are two categories of general purpose tool grinders: dry and wet grinders.

1. *Dry Tool Grinders.* Dry tool grinders are equipped with a coarse grit wheel and are used primarily for rough grinding. They are often used to machine tool steel blanks into a desired shape. Because of the coarseness of the wheel, overheating can be a problem.

2. *Wet Tool Grinders.* Wet tool grinders have a finer grit wheel and, as the name implies, use water or soluble oil solution to control overheating. These grinders are used to resharpen tools or to finish rough ground tooling.

Precision Grinding Machines. Precision grinding machines are used to machine parts to close tolerances and produce a high quality surface finish. Many parts can be easily ground to tolerances within 0.0001 inch. There are three types of precision grinding machines found in the maintenance shop: cutter and tool grinders, cylindrical grinders, and surface grinders.

1. *Cutter and Tool Grinders.* These grinders are specifically designed for the precision sharpening of cutting tools such as milling cutters. Though considered somewhat of a specialty tool and not found in all shops, cutter and tool grinders are capable of performing various light duty operations such as surface and cylindrical grinding.

2. *Cylindrical Grinders.* Also referred to as *universal grinders,* cylindrical grinders are used to machine the internal and external surfaces of cylindrical parts. Designed with a headstock assembly and a tailstock assembly, they resemble a lathe. As the workpiece rotates between the two assemblies a rotating grinding wheel is fed into the part's surface.

3. *Surface Grinders.* Perhaps the most common type of precision grinding machine found in maintenance shops is the surface grinder. Used for the grinding of flat surfaces, the surface grinder is equipped with an oscillating table that has a magnetic chuck for holding the work. The chuck and workpiece are then fed under the grinding wheel by manual or automatic feed. The grinding wheel spindle can be either horizontal or vertical to the table. (See Figure 3-3.)

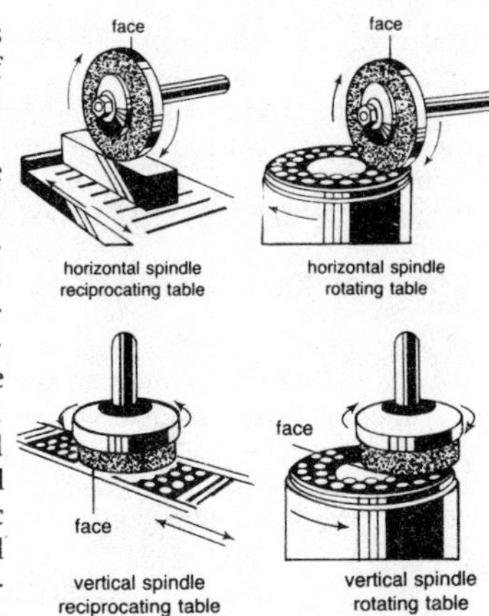

Figure 3-3. Basic types of surface grinding.

Use of Grinding Machines

1. Make sure that the grinding wheel is in good condition, is the proper type for the work to be done, and is properly and securely installed.
2. Use a grinder only when wheel guards are in place. Wheel guards protect the operator should the wheel break or shatter.
3. Never touch a grinding wheel with your hands or foreign objects.
4. Make sure that the work is firmly secured before grinding begins. Never hold the workpiece with a cloth and do not wear gloves.
5. Grind on the face of the wheel.
6. Always stop the grinder and make sure that the grinding wheel has stopped completely before making any adjustments.
7. Avoid grinding consistently in the same spot on the wheel.
8. There should be only one operator at a time working on the grinding machine.

MILLING MACHINES

The milling machine is primarily a metal-cutting machine that is powered to produce surface contours by a rotating cutter. The multiple tooth cutter is called a *milling cutter*. Most milling machines are equipped with both a manual and automatic feed for longitudinal and transverse (crossfeed) axes.

TYPES OF MILLING CUTTERS

Milling cutters come in a wide variety of designs, sizes, and construction. There is no one standard method by which cutters are classified. A listing of several common classification methods is presented in Table 3-4. For discussion purposes, milling cutters are grouped here into three major categories: arbor-mounted, end milling, and face milling cutters.

TABLE 3–4
Common Classification Methods

Classification Criteria	Examples of Cutters
Application or use	Keyset, T-slot, and gear cutters
Construction	Solid, carbide-tipped, and inserted blade cutters
Mounting method	Arbor, straight shank, and taper shank cutters
Relief of teeth	Profile and formed cutters

Arbor-Mounted Milling Cutters. Arbor-mounted milling cutters are designed with a machined hole and keyway for mounting on a milling machine arbor. The hole is precision machined, and the keyway is used for locking onto the arbor key. Arbor-mounted milling cutters have cutting edges located around the periphery and sometimes along the sides.

A number of different types of arbor-mounted milling cutters have been designed. These include angle, forming, metal slitting, side, and plain milling cutters.

1. *Angle Milling Cutters.* Angle milling cutters are made with cutting edges set at a specified angle. They may be either single or double angled. Standard single-angle cutters will machine an angular surface at common angles such as 45°, 60°, and 90°. Double-angle cutters form a V-shaped slot in the workpiece at similar standard angles.
2. *Forming Milling Cutters.* Forming milling cutters are used to conform to a given contour or profile. There are two types of forming cutters: *profile ground* and *formed relieved* cutters. Profile ground cutters have cutting edges similar to plain milling cutters; formed relieved cutters have cutting edges that are cut by a relieving lathe attachment.
3. *Metal Slitting Saws.* Saw cutters are used to cut narrow slots into the workpiece. They can have cutting edges on their sides as well as on the periphery. Common saw widths range from 1/32 to 3/16 inch.
4. *Plain Milling Cutters.* Also known as *slab milling cutters*, plain milling cutters are used to machine flat or plane surfaces.
5. *Side Milling Cutters.* Side milling cutters are similar to plain milling cutters, but they have additional cutting edges along the sides. Side milling cutters are considered a general purpose cutter that can be used for a variety of cutting operations, including side, straddle, slot, and plane milling.

End Milling Cutters. A significant number of milling cutters used are of the end milling type. These cutting tools are designed with cutting edges on the end surface as well as along the periphery. End milling cutters are held in the milling machine spindle by a *collet chuck* or a similar holding mechanism. This type of cutter is considered the most adaptable and versatile of all milling cutters because it can be used for a variety of operations including the milling of flat and plane surfaces, slots, profiles, contours, and holes.

End mill cutters are available as either right- or left-hand cutters. Right-hand cutters can only cut when rotating in a counterclockwise direction, while the left-hand cutter machines in a clockwise rotation.

There are four general types of end milling cutters: center cutting, rough cutting, shell end, and two fluted end mills.

1. *Center-Cutting End Mills.* Center-cutting end mills have two cutting surfaces that come together at the end of the cutter and give the appearance of a twist drill. The center-cutting end mill is used to machine holes and openings into workpieces.
2. *Rough-Cutting End Mills.* Rough-cutting end mills are used for rough milling where large amounts of stock must be removed rapidly.
3. *Shell End Mills.* Shell end mills are used primarily for surface and corner machining cuts (e.g. square, round, or chamfer shaped) and are rarely used for slot cutting. Their design is a compromise between face milling and end milling cutters.
4. *Two-Fluted End Mills.* Also known as *two-lipped end mills*, two-fluted end mills can be easily and directly fed into the workpiece. They are frequently used to machine complex contours.

Face Milling Cutters. The majority of flat or plane surface milling is accomplished with face milling cutters. Though these cutters are available in sizes that exceed 6 inches in diameter, only the smaller sizes are useful in maintenance and repair work. Since the majority of face milling cutters are used for high production operations, they are seldom encountered within the trade.

TYPES OF MILLING MACHINES

There are a variety of milling machines for high quality and single-lot production. Of primary concern to the machine and maintenance mechanic are those milling machines that are best suited for single-lot jobs. Of all the metal machines, they are one of the most versatile tools available.

There are four broad categories of milling machines: knee and column, fixed bed, planner type, and special milling machines. The first, however, is best suited for maintenance type work. The other three are more applicable to high production situations.

Knee and Column. The knee and column milling machine is an extremely versatile tool, capable of performing a large number of machining operations, including drilling, reaming, boring, planing, and contouring. These machines are noted for their ability to adjust the workpiece in three directional movements that correspond to *coordinate axes*. These are: x-axis, or left to right movement; y-axis, or front to back movement; and z-axis, or up and down movement.

The milling machine is equipped with a table that moves along the x and y axes and is mounted on a *saddle*. The saddle movement allows movement along the z-axis; this last directional movement occurs when the knee travels vertically along the column.

The knee and column milling machine is further divided into three categories that are described here.

1. *Plain Knee and Column Milling Machines.* This type of milling machine, which is used in machine and maintenance shops, job shops, tool and die shops, and other small commercial operations, can perform all common milling operations. Special operations, such as helical milling, can also be accomplished with special attachments.

   Workpiece movement is provided as the knee slides vertically along the column face and is power fed except for light duty work. The transverse feed is accomplished when the saddle slides along the knee. Longitudinal feed occurs as the table moves over the upper slide of the saddle and 90° to the spindle's axis. Micrometer dials are located along the longitudinal crank or wheel for accurate settings.

   T-slots are milled into the top of the table so that the workpiece can be fastened by anchor bolts. These are also used to hold down machine vises and other attachments. A power feed is used to move the workpiece across the cutting tool which is located above the work. A selected gear change box provides different rates of feed which are given in inches per minute of table travel. Automatic feed is available in all three directions.

2. *Universal Knee and Column Milling Machines.* In milling machines of this type the table housing can be swiveled. This allows the table to be rotated at any angle to the cutting tool or spindle axis. This feature makes the universal milling machine much more versatile than the plain knee and column milling

machine. Feeds and axes movements, however, are basically the same as the plain milling machine.

3. *Vertical Knee and Column Milling Machine.* Of all the milling machines described here, the vertical knee and column type is perhaps the most common type found in maintenance shops. A light duty machine, it has great flexibility in various types of machining.

The primary difference between vertical milling machines and the plain and universal types is that the machine's spindle is set vertically to the table—parallel to the face of the column. This milling machine is best suited for using end mills and face milling cutters. Common operations performed on this mill are the machining of contours, planing surfaces, and the locating and machining of holes and cavities. (See Figure 3-4.)

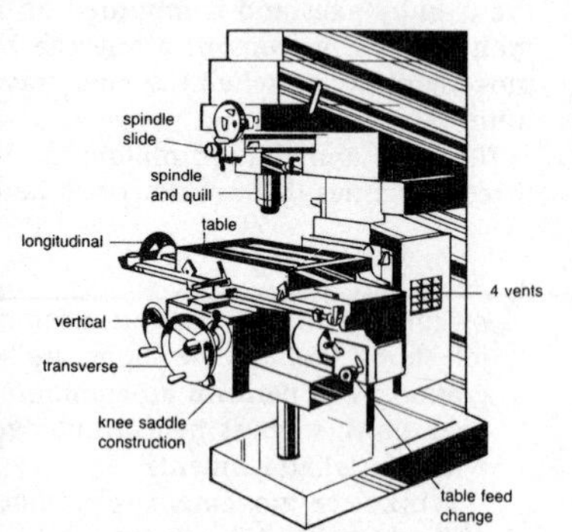

Figure 3-4. Vertical milling machine.

USE OF MILLING MACHINES

There are two milling forms that must be taken into consideration when discussing the proper use of mills: horizontal milling and vertical milling. One of the simplest, and perhaps most common, processes performed on milling machines is the milling of flat surfaces. The procedures used here are also applicable to other milling processes.

Horizontal Milling

1. Clean the top of the table, and mount the vise on it.
2. Make sure that the jaws of the vise are either at a right angle to the face of the column or exactly parallel to it.

3. Place parallels in the vise, and mount the workpiece on them. Tap the workpiece with a mallet to seat it.
4. Mount the arbor in the spindle.
5. Select the proper cutter, place it on the arbor, and key it to the arbor.
6. Swing the overarm into place and lock it.
7. Adjust the correct speed and feed.
8. Position the workpiece under the rotating cutter until it just touches the surface, and set the micrometer dial to 0 on the index line. Back the workpiece away from the cutter, and raise the table for the depth of cut desired.
9. Tighten all locks and turn on the coolant and power feed. Do not stop the feed during the machining operation.

Vertical Milling

1. Select the end mill and mount it in the spindle. Check to see that the spindle is perpendicular to the table.
2. Set the workpiece in the vise and secure it.
3. Adjust the machine for proper speed and feed.
4. Check the length of cut to be taken.
5. Move the workpiece under the cutter and start the machine.
6. Raise the knee until the cutter just touches the workpiece. Set the micrometer dial to 0 on the index line.
7. Move the table so that the workpiece just clears the cutter, and raise the knee to the required height.
8. Move the table by hand until the cut is started, and then turn on the power feed.
9. After the cut has been taken, measure the workpiece and take a second cut if necessary.

SANDERS

Sanding operations are similar to grinding—both are abrading processes. Various minerals are used as abrasive agents. Three common natural minerals are garnet, emery, and quartz (flint).

In addition, aluminum oxide and silicon carbide are also frequently used.

ABRASIVE PAPERS AND BELTS

Abrasive minerals are crushed and graded for making products such as papers and belts. These products come in an open coat in which the abrasive grain is evenly distributed over 50 to 70 percent of the surface. Closed coat abrasives are those in which the abrasive grain covers the entire surface of a material.

Table 3-5 lists the comparative grit numbers for common abrasive papers, discs, and belts. It should be noted that because of the variation in hardness, a coarser grade of soft abrasive will be required to do the same job as can be done with a harder mineral.

TYPES OF POWER SANDERS

There are two basic types of sanders used in the shop—belt sanders and disc sanders. There are also some sanders that are equipped with both disc and belt attachments and are known as combination sanders; these sanders are capable of machining any material as long as the appropriate abrading medium is used. For example, silicon carbide and aluminum oxide are used with steel and steel alloys, while garnet is widely used for sanding wood.

Belt Sander. The belt sander is usually equipped with a tilting table that can be adjusted between 90° and 45° to the belt. The table is designed with a slot parallel to the sanding surface for a miter gage. This type of sander provides for a particularly uniform cutting speed across the entire surface of the belt that results in a smooth and even cut.

Disc Sander. Most disc sanders have an adjustable tilting table that is similar to that found on belt sanders. The primary difference between the two types of sanders is that the disc sander has a rotating abrasive disc that is used to remove the material and is held in position by means of an adhesive compound.

TABLE 3–5
Approximate Comparison of Abrasive Grit Numbers

Aluminum Oxide and Silicon Carbide	Garnet	Flint	Grade
400-10/0	—	—	
360	—	—	
320-9/0	—	7/0	Very Fine
280-8/0	8/0	6/0	
240-7/0	7/0	5/0	
220-6/0	6/0	4/0	
—	—	3/0	
180-5/0	5/0	—	
150-4/0	4/0	—	
—	—	2/0	Fine
120-3/0	3/0	—	
—	—	0	
100-2/0	2/0	—	
—		1/2	
80-0	—	—	
—	—	1	Medium
60-1/2	1/2	—	
50-1	1	1 1/2	
—	—	2	
40-1 1/2	1 1/2	—	
—	—	2 1/2	Coarse
36-2	2	—	
30-2 1/2	2 1/2	3	
24-3	3	—	
20-3 1/2	3 1/2	—	Very coarse
16-4	—	—	
12-4 1/2	—	—	

Use of Belt and Disc Sanders

1. Check to see that the sanding disc or belt is in good condition, of the proper grit size for the work to be done, and is properly aligned and secured.

2. Make certain that the stock being sanded is resting firmly against the back stop on the bed of the machine. Make all adjustments while the power is off.
3. Set up the sander to take as small a cut as possible.
4. Keep your full attention focused on the work. Small pieces of stock should not be held in the hands. On thin stock use a push block.
5. If sanding wood, sand in the same direction as the wood grain. When sanding metal, do not apply excessive pressure that will cause overheating, and potential loss of temper.
6. Use successively finer grit until the desired finish is obtained.

SAWS

Power saws are invaluable because they are capable of cutting through a variety of different materials with speed and accuracy. Some saws are designed specifically for use with softer material, such as wood, while other saws are used primarily for metal cutting.

TYPES OF SAWS

Band Saw. A band saw has a continuous blade that travels around rotating wheels. The blades are made of high carbon steel with a flexible back and hardened teeth. They come in a variety of widths and pitches. Common widths are 1/4 inch (6 mm), 3/8 inch (10 mm), 1/2 inch (13 mm), 5/8 inch (19 mm), 3/4 inch (19 mm), and 1 inch (25 mm) inch. Pitches available are 6, 8, 14, and 18 teeth per inch.

1. *Horizontal Band Saw.* The horizontal band saw is used primarily for metal cutting where the saw blade travels in a horizontal plane or a plane slightly inclined from the horizontal. These saws are used to cut off stock square or at an angle. The feed of these machines is usually hydraulically operated and comes with an adjustable vice and stock stop. Most horizontal band saws have different cutting speeds.

2. *Vertical Band Saw.* The vertical band saw is widely used to cut metal, wood, and other materials. Typically employed for cutting curved surfaces and for contour cutting, the vertical band saw can also be used to make straight cuts. Contour sawing is a fast, accurate, and efficient method for producing intricate curved or irregular cuts in almost any materials. Radii as small as 1/16 inch (1.6 millimeters) can be cut. Both internal and external contours can be sawed on the band saw.

Circular Sawing Machines. Circular sawing machines are used as a cutoff method for wood, metals, and many plastics. However, most circular sawing machines found in the maintenance shop are limited to those used for cutting wood. Metal circular saws are considered specialized machines more applicable to tool and manufacturing situations.

1. *Radial Saw.* The radial saw is a versatile power tool used in all types of maintenance construction and repair jobs. What makes this machine different from many saws is that the material being cut always remains stationary, while the saw moves. This is often desirable when making complicated cuts, because the operator is able to see the cut at all times.

   The saw is mounted to a turret arm that allows it to swing in a full circle about the horizontal plane, while keeping the saw over the table. The motor unit is directly attached to the saw, so that it will also tilt and move. This flexibility is advantageous, especially when a dado cut is being made, including left-hand mitering. The position of the fence can be changed, and as a safety feature, all controls are positioned out of the way of the saw. (See Figure 3-5.)

2. *Table Saw.* The table saw is one of the most frequently used power tools for woodworking. It uses a circular saw blade up to 16 inches in diameter and is equipped with many attachments and safety devices. A handwheel raises or lowers the blade, as required for various stock thicknesses, and a tilt handwheel allows the operator to change the angle of the blade.

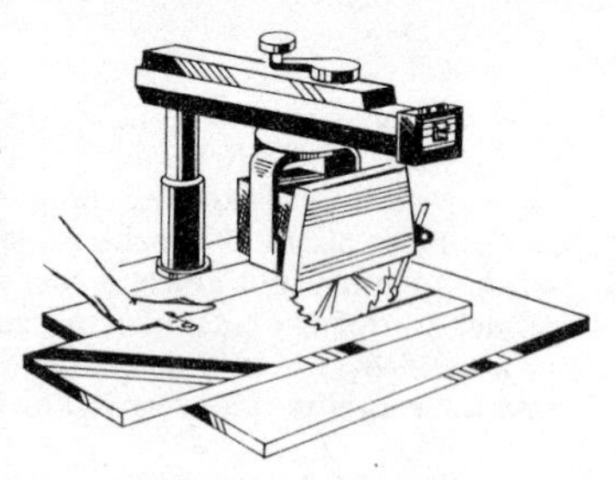
Figure 3-5. Radial arm saw.

The saw blade is protected by a blade guard that usually incorporates an anti-kickback device. The controls of the ripping fence are positioned forward and near the operator as a safety feature. This fence locks to two guide bars located at the front and back of the machine and allows for various positioning adjustments.

It should be noted that the saw blades used on the table saw are the same as found on the radial saw. The common ones used for both are the crosscut, rip, combination, and hollow ground blades.

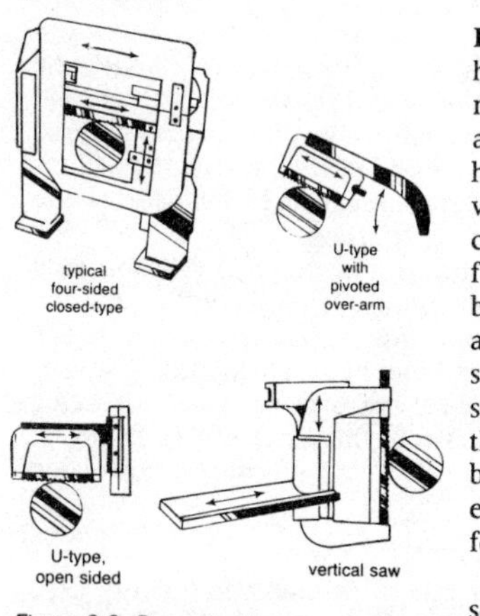

Figure 3-6. Power hacksaw arrangements.

Power Hacksaw. Power hacksaws are used for cutting metal stock to various lengths at fairly close tolerances. The hacksaw is equipped with a vise that can be swiveled to cut stock at an angle. A saw frame is used to hold the blade. As the frames moves in a back-and-forth motion, sawing occurs on the forward stroke. The return stroke lifts the frame slightly to minimize blade wear. Most saws are equipped with a hydraulic feed system. (See Figure 3-6.)

Power hacksaw blades are somewhat different from hand hacksaws. They are heavier and have a coarser tooth pitch. They are available in ten different lengths ranging from 12 inches (305 mm) to 36 inches (914 mm). Width range is 5/8 inch (16 mm) to 4 1/2 inches (114 mm). Common pitches of teeth are 2 1/2, 3, 4, 6, 10, 14, or 18 teeth per inch. Examples of blade selection for common types of cutting are:

1. *Light angle iron, pipe, and rods:* light to moderate feed pressure, 1-inch blade width with a thickness of 0.050 inch.
2. *General purpose cutting:* moderate feed pressure, 1 1/4- to 1 1/2-inch blade with a thickness range of 0.062 to 0.075 inch.
3. *Large sections:* heavy feed pressure, 1 3/4- to 2-inch blade width with a thickness range of 0.088 to 0.100 inch.

LATHES—TURNING

Turning includes a variety of operations performed on lathes. On a lathe, the workpiece is rotated while the tool is reciprocated into the work to complete a given sequence of cuts. Examples of common machining operations performed on lathes include facing, cylindrical turning, shoulder turning, threading, boring, drilling, mandrel turning, knurling, and taper cutting. (See Figure 3-7.)

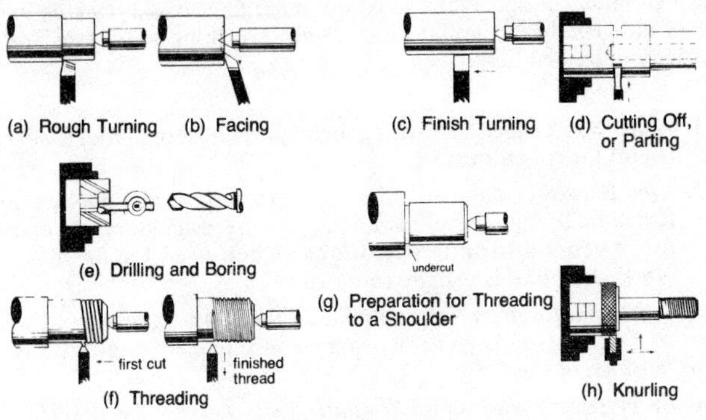

Figure 3-7. Common turning operations.

There are three basic types of lathes: metal, wood, and metal spinning. Of these, the metal lathe is the most frequently used. Wood lathes are seldom found in the maintenance shop, and the metal spinning lathe is almost never used by the mechanic. In many machine and maintenance shops, nonmetallic materials, such as wood and plastics, are also machined on the metal lathe. In light of this situation, our discussion will be limited to the metal lathe.

TYPES OF CUTTING TOOLS

There is a wide variety of cutting tool designs and materials used for lathe work, ranging from standard tool steels to industrial diamonds. When selecting a specific type and material for cutting tools, there are two factors that are always considered: type of work to be performed and the quality of the final product.

Cutting Tool Designs. Critical to the lathe tool bit are the clearance angles which allow the part of the tool bit directly under the cutting edge to clear the work while making a chip. Too much clearance will weaken the cutting edge and make the tool chatter during turning.

Five basic and standard lathe tool bit forms are used for most turning operations. It is important to note that mechanics are required to grind their own tool bits from tool bit blanks. Thus, the dimensional considerations given are important for the making of tool bits. The basic tool bit forms used for turning operations are briefly described here.

1. *Round Nose Cutting Tool.* This general purpose cutting tool is also useful for rough turning.
2. *Right-Handed Cutting Tool.* This is an excellent turning tool used for shouldering operations toward the headstock, facing, and other general turning operations. When used for finishing work, the point is usually rounded.
3. *Left-Handed Cutting Tool.* Similar to the right-handed cutting tool, this tool is for general turning and shouldering toward the tailstock of the lathe.
4. *Heavy Duty Right-Handed Toughing Tool.* This tool is used for taking deep cuts toward the headstock of the lathe. The clear and rake angles of the tool should be reversed for left-handed turning.
5. *Right-Handed 60° V-Threading Tool.* Also used for cutting toward the lathe's headstock, this tool is used for thread cutting. The side clearance angle of this tool should be reversed for left-handed threading.

Cutting Tool Materials. There are seven broad categories of cutting tool materials available for lathe work.

1. *Carbon Tool Steels.* These cutting tool materials have been in use longer than other tool materials. They are inexpensive and heat treatable, and have good cutting properties. They are suited for most general purpose turning operations.
2. *Cast Alloys.* Designed to operate at high temperatures, cast alloys have excellent wear resistance. They are used primarily for deep rough cutting operations and do not require the use of cutting fluids.
3. *Cemented Carbides.* Primarily found in production settings, these cutting tools are also referred to as *throw-away tips* because they are replaced rather than resharpened. They have excellent cutting properties and maintain their hardness over a wide range of temperatures.
4. *Ceramic Tooling.* Cutting tools made from nonmetallic materials are known as *ceramic* or *oxide cutting tools.* Commonly made from aluminum oxide, these tools are used for machining at high speeds and at high temperatures.
5. *High Speed Steels.* High speed steels are common cutting tool materials. Best suited for production settings, they are used in a wide variety of turning operations. High speed steels maintain hardness and strength at high operating temperatures.
6. *Industrial Diamonds.* Also known as *borts,* industrial diamond cutters are used when dimensional accuracy and fine surface finishing are critical.
7. *Medium Alloy Tool Steels.* These cutting tools are best suited for operations that do not involve high speed turning (e.g., boring and threading). They demonstrate a greater resistance to wear than typical carbon steels.

TYPES OF LATHES

Lathes can be classified by two different methods. In the first, the size or capacity of the lathe is described in terms of *swing* and the *length of bed.* Swing pertains to the maximum diameter of the workpiece that can be rotated; the length of bed pertains to the maximum workpiece length that can be mounted between the headstock and tailstock.

The second classification method is based upon application and is broken down into five groups: bench, engine, turret, production, and vertical lathes. The last three are equipped for producing large quantities of duplicate parts at high production rates and are primarily used for manufacturing. Bench and engine lathes are particularly well suited for the maintenance shop.

Bench Lathe. The bench lathe is commonly found in small maintenance shops, where parts are individually made or repaired. It is the smallest type of lathe and can easily be mounted on top of a bench—hence the name. Bench lathes are designed for precision machining of small parts, and accuracy depends on the competence of the operator. (See Figure 3-8.)

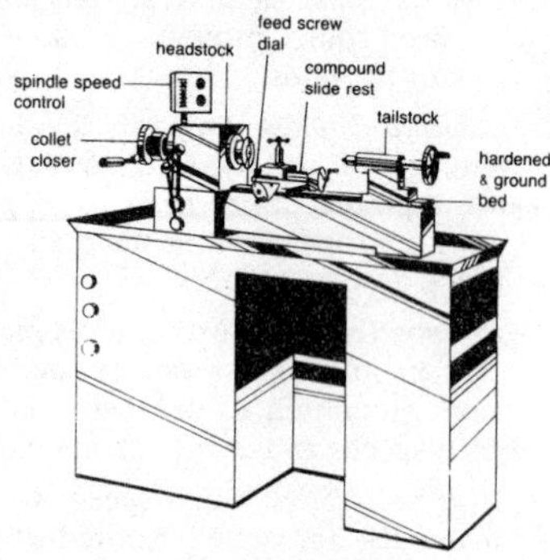

Figure 3-8. Bench lathe.

Except for the power drive, most bench lathes are manually operated. Longitudinal and cross feeds are controlled by graduated indicators. Some bench lathes are equipped with sophisticated components that can control machine speeds and feeds. Attachments for drilling, boring, threading, grinding, and other operations are available.

Engine Lathe. The most widely used and fully powered lathe found in the machine shop today is the engine lathe. The term "engine" lathe refers to the early practice of driving these lathes with a steam engine. Today, these lathes are driven by electric motors.

The engine lathe is the most popular type of lathe because of its simple construction, versatility, and easy maintenance. Examples of operations that can be performed on these machines are cylindrical turning, flat surface facing, drilling, boring, threading, internal hole finishing, and knurling. As with the bench lathe, the accuracy of the part produced depends on the competence of the lathe operator.

USE OF LATHES

Care and accuracy is paramount in lathe work. To describe the procedures that should be followed in all lathe operations would be beyond the scope of this book. However, to provide apprentice mechanics with a reference point, a description of the procedures followed in turning between centers will be given here.

Facing Workpiece

1. Cut the stock to desired length.
2. Mount a universal, three-jaw chuck onto the headstock spindle.
3. Insert the workpiece into the chuck and tighten it in place.
4. Insert a right-hand cutting tool into the tool holder, and adjust the toolholder so that the cutting edge of the tool is at an approximate 90° angle to the spindle.
5. Square up the end by moving the carriage to the left so that it just makes contact with the workpiece at the center point.
6. Move the crossfeed toward you. The first cut will be a rough cut from the outside toward the center. For the finishing cut, feed the crossfeed from the center to the outside of the workpiece.
7. Reverse the workpiece and face the other end.

Rough Turning

1. Insert a center drill into a chuck, and insert it into the tailstock.
2. With the workpiece in the three-jaw chuck, drill a center hole in the end of the workpiece. The depth of the drill should be approximately two-thirds the length of the center hole drill taper, thus forming a countersink.

3. Remove the workpiece, turn it around, and drill another center hole as done in step 2.
4. Remove the three-jaw chuck, and replace it with a faceplate.
5. Fasten a dog to the workpiece.
6. Insert a live-center in the spindle, and hold the workpiece as you insert its countersunk center hole onto the live-center.
7. With a dead-center mounted into the tailstock, move the tailstock forward until it just enters the center hole at the other end of the workpiece and lock it in position.
8. Lubricate the dead-center end of the workpiece, and hand tighten firmly, but not too forcefully, and lock the dead-center in position.
9. Set the tool on center, and turn it slightly away from the headstock.
10. Set the lathe for correct speed and feed. Move the carriage towards the headstock to make sure that it can travel the required distance without striking the compound rest or lathe dog.
11. Determine the depth of roughing cut. Adjust the micrometer collar on the cross-feed to zero and lock in position.
12. Back the tool off and move the carriage to clear the end of the workpiece to the right, and turn the cross-feed in for the desired cut.
13. Turn on the power, engage the carriage clutch, and make a trial cut to make sure that the diameter desired is true.
14. Complete the rough cut over the length of the workpiece.

Finish Turning

1. Prior to any finish turning, check the diameter of the workpiece to find out the amount of stock to finish turn to the specified dimensions.
2. Insert a right-hand finishing tool in the tool holder, and be sure that it is sharp.
3. Turn the power on and move the tool so that it just touches the workpiece, and then turn off the power.

4. Adjust the cross-feed collar to zero, and set the feed in for the required depth.
5. Take a test cut to assure that the diameter turned is what is desired.
6. Turn on the power and turn the remainder of the workpiece.

PORTABLE POWER TOOLS

The major advantage of portable power tools over stationary tools is their greater flexibility. Portable tools can be taken to the job site to reduce work time and provide for greater efficiency.

Portable tools can be powered by air (pneumatic) or electricity. Electric power tools are the most common, but, over the last 10 years, there has been a significant increase in the use of pneumatic tools. Since the function of portable power tools is similar to that of stationary tools, this section only describes the major types of portable power tools used by the mechanic.

DRILLS

The portable electric drill is perhaps the portable power tool most frequently used by machine and maintenance mechanics. This tool comes with either a pistol grip or D-shaped handle for standard drills and a spade handle for heavy-duty drills. Some are also available with an auxiliary handle for better control. Portable drills are used to drill holes into a variety of materials including wood, metal, ceramic tile, concrete, and brick. (See Figures 3-9 and 3-10.)

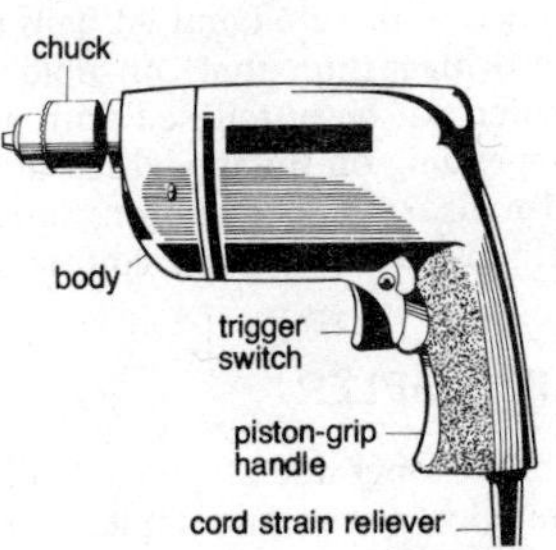

Figure 3-9. Pistol grip power drill.

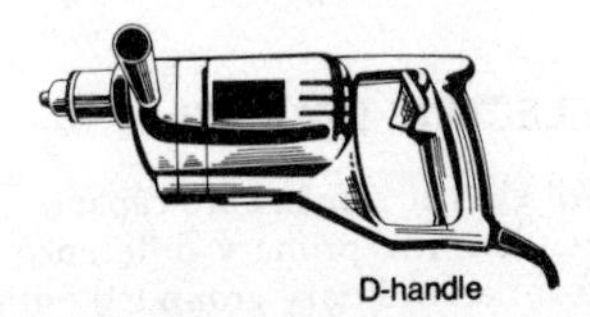

Figure 3-10. D-handle power hand drill.

Portable drills are available in sizes that can accept drills having a maximum diameter range between 1/4 inch and 1 1/4 inches. Drill speeds on some tools can be varied, as can the drilling direction (that is, clockwise and counterclockwise rotation). Battery operated drills are also available, with some having a rechargeable battery in the handle of the drill. However, most power drills come with a cord that can run from a 115 volt outlet.

NAILERS AND STAPLERS

The majority of staplers and nailers on the market today are pneumatic. Air-operated and electric staplers and nailers come in different sizes and shapes and are designed for different uses.

AIR NAILER

Pneumatic nailers are used for heavy nailing jobs and are designed to drive 6d and 9d nails in hardwood. Each nailer is loaded with magazines that can hold up to 300 nails. The air-operated nailer can be purchased with a trigger or touch-trip mechanism. Depending on the model selected, it is possible to use various types of nails, such as common, finish, brad, or T-nails. Some designs will also automatically countersink nails.

AIR STAPLER

The air-operated stapler is quite versatile. In addition to roofing shingles, pneumatic staplers can also be used to install insulation, ceiling tile, vapor barriers, building paper, metal lath, plywood, and subflooring.

ELECTRIC STAPLER

All electric staplers are capable of doing the same work as the air stapler. The primary difference is its power source—it operates from an ordinary grounded outlet instead of an air compressor.

PLANES AND ROUTERS

The portable power plane can provide fast, accurate edging for various cabinetworking operations, as in the fitting of doors, drawers, window sash, storm sash, screens, shutters, transoms, and trim. Because of the speed at which it accomplishes its job, it is significantly better than the hand plane.

The majority of power planes come with a spiral cutter that produces a fine, smooth finish. These planes are capable of planing surfaces up to 2 1/2 inches wide. A graduated dial on the front of the machine is used to adjust the depth of cut. The maximum depth of cut found for these tools is 3/16 inch. The angle fence of the tool can be tilted to a maximum angle of about 15° outboard and 45° inboard.

PORTABLE ROUTER

The portable router is based upon the planing concept. This tool is extensively used in shops engaged in construction, fine joinery, inlay work, and other wood finishing. Router cutters come in a variety of designs to accomplish jobs such as beading, grooving, fluting, rounding, mortising, and dovetailing. Standard router equipment usually includes a combination straight and circular guide, slot and circle cutting attachment, template guides, and joinery templates.

SANDERS

There are three basic types of portable sanders available to the mechanic: belt, disc, and orbital. The belt sander is primarily used for woodworking; the disc and orbital sanders are used for woodworking, plastics, and metal work.

BELT SANDER

The belt sander is primarily used for sanding large flat areas, as in the removal of old paint and varnish for refinishing. Sanding belts come in coarse, medium, and fine grades, and have a directional marking on the backing (this shows the direction the belt should face when mounted). Most belt sanders come with a dust bag that collects the generated wood dust.

DISC SANDER

The primary advantage of the disc sander is its ability to sand uneven and curved surfaces. It is recommended that this type of sander not be used on untreated wood, but it can be used for operations on plastics and metal. The disc sander is considered more flexible than the other two types of portable sanders because of the variety of accessories that can be used in place of the sanding disc. For example, wire torque brushes are used for cleaning and removing paint and other deposits; an abrasive wheel is used for grinding operations, and a felt pad mounted on a rubber pad can be used for polishing operations.

ORBITAL SANDER

The orbital sander, also known as a *finishing sander*, is used for small and less accessible surfaces. Unlike the other sanders, the orbital sander uses standard sheet abrasive paper and cloth as well as paper that can be hand cut to size. As its name indicates, the sander oscillates in an orbital pattern and is used for finish sanding operations.

SAWS

There are three types of portable power saws that are commonly used within the trade. These are the portable power saw, sabre saw, and reciprocating saw.

PORTABLE POWER SAW (PORTABLE CIRCULAR SAW)

The portable power tool, also known as a portable circular saw, is primarily used on construction and maintenance work. It is a hand-held saw and has a circular blade available in diameters from 6 to 22 inches, the most common sizes being 6 1/2 inches and 7 1/4 inches. The diameter of the blade dictates the maximum depth of cut.

The portable power saw is used for crosscutting and ripping wood as well as for cutting soft material such as plastics and particle board. The base of the saw can be raised or lowered to adjust the depth of cut. Most saws are capable of making a bevel cut up to 45°.

RECIPROCATING SAW

This saw is used for cutting wood, plastic, and metal, depending upon the type of blade used. The reciprocating power saw is a general purpose saw and derives its name from its reciprocating (up and down) cutting action. The cutting stroke of the saw is about one inch, and most models have a variety of cutting speeds. This saw is capable of starting its own hole.

SABRE SAW

The sabre saw is sometimes known as a *bayonet saw* because of the appearance of the protruding saw blade. It is a versatile saw and can be used for cutting a wide variety of materials and products.

Sabre saws usually have an orbital blade saw with a stroke ranging from 7/16 inch to 1 inch. The tip of the blade is pointed, which means that it is also capable of starting its own hole. Its base, known as the *shoe,* can be adjusted to make bevel cuts. Perhaps the best feature of this tool is its ability to cut sharp angles and curves.

PART TWO

PREVENTIVE MAINTENANCE AND TROUBLESHOOTING

4

Buildings and Plants

One of the more common responsibilities of the mechanic is plant or building maintenance. Organizing a standardized maintenance program with clearly defined responsibilities and comprehensive work schedules is important. The maintenance mechanic also needs to know how to troubleshoot and recognize problems different from those in the normal work schedule. This chapter presents preventive maintenance checklists and troubleshooting procedures commonly used in maintaining buildings and physical plants.

The first step in developing a preventive maintenance program is to make a complete list of all maintenance jobs that have to be performed in each area of the plant. The list should include how often each job should be performed and an estimate of the time required for each job. The checklists presented in this section are applicable to most facilities, although each facility has unique design and maintenance requirements. Similarly, the troubleshooting tables presented will help in the diagnosis of many problems, but they may not pinpoint concerns in a particular facility.

ASPHALT PAVEMENT

Asphalt pavements and walkways are easy to maintain, especially if they are made of asphaltic concrete or asphalt macadam. Asphalt surfaces are frequently renewed by a light seal coat that produces a smooth waterproof surface. Depending on the traffic and climate, surface treatment need be applied only every three to five years, and heavy-duty asphalt surfaces should not require treatment for at least ten years.

BOILERS

Inspection schedules for boilers and pressure vessels are normally dictated by state or local inspectors that have jurisdiction over the facility. In some cases the insurance company that provides coverage for a company also requires inspections at periodic intervals. The inspections should be scheduled at a time convenient for both the inspector and the chief maintenance mechanic (or chief engineer).

Table 4-1 lists a number of maintenance practices that should be followed to ensure efficient and safe boiler operation.

TABLE 4–1
Boiler Maintenance

Sequence of Operations	Description
Blowing tubes	Before taking a boiler out of service, remove soot from the fireside by blowing air or steam against the tubes with an automatic soot blower or hand lance. Never blow tubes with wet steam.
Boiler cleaning	Remove and clean upper drum manhole covers by hosing with warm water. Use a heavy water stream to clean the drums and tubes. Then, open furnace access doors and settings and clean with available cleaning tools.

TABLE 4–1 (*continued*)

Sequence of Operations	Description
Fusible plug cleaning	Remove cross tee plugs in water column to inspect the piping. Clean piping with a brush and replace the fusible plug once a year.
Stack cleaning	Inspect breeching, all dampers, and damper operating mechanisms.
Valve packing	Check that all valves are packed and in full operating condition.
Piping inspection	Inspect all water piping, pressure connections to gages, automatic control lines, and other piping. Also inspect feedwater regulating equipment and all piping and supports. Check valves and control valves.

COMMON PROBLEMS

The mechanic should be aware of common boiler problems and should be able to troubleshoot and identify specific problems. (See Table 4-2.)

Abrasion. The wearing away of metal within a boiler system is a common problem. Causes of boiler abrasion include the movement between parts in contact; fly ash concentrating and striking the boiler's surface; fuel, water, or steam flow rubbing against headers or tubes; and frequent use of cleaning tools.

Corrosion. The major corrosion problem associated with boilers is rusting. There are two basic causes of boiler corrosion. The first, which normally occurs fireside, is acid attack on the metal caused by moisture accumulation of soot or fly ash within the boiler and/or setting. The second is washing the boiler and then allowing it to stand.

Leakage. Leakage at the rolled joints inside the boiler is a significant problem that can be detected by the presence of moisture or a flow of water when the boiler is subjected to hydrostatic testing. Two other indicators of joint leakage are the accumulation of chemicals or cemented soot deposits and a hissing sound when the boiler is under steam pressure.

TABLE 4–2
Troubleshooting Boiler Problems

Problem	Possible Causes
Hot boiler metal	Metal damage due to scaling.
Sludge	Impurities in the boiler water or improperly treated water.
Stoppage of circulation through drums, headers, or supply tube	Excessive sludge formation.
Improper circulation	Failure of tubes.
False water level readings	Defective gage or leakage at connections or at any other valves or fittings in the gage.
Low water level in boiler	Failure of feed pump. Leaks in the discharge relief line of the feed pump. Improper regulation of defective valves. Low water in feed tank. Water at pump inlet too hot. Leakage in the feedwater heater. Leakage in the boiler itself. Priming.
Priming	Forcing the boiler beyond its rated capacity. Increasing load too rapidly.
Excessive foaming	Abnormal agitation of the water level. Presence of contaminated water.
Sudden increase in load	Excessive foaming.

CLEANING

The advantages of a good cleaning program should not be underestimated. Lack of cleanliness reduces machine efficiency and increases wear and tear, thereby increasing the chance of breakdown. In fact, it has been found that in some cases an effective cleaning program can increase productivity by 300 percent.

Table 4-3 gives a recommended cleaning schedule for a typical commercial building.

TABLE 4–3
Cleaning Schedule

Area of the Building	Cleaning Jobs	Frequency
Corridors	Clean sand urns	4 times a day
	Sweep floor	Once a day
	Damp mop floor tile	Once a day
	Damp mop terrazzo	Once a day
	Damp wipe ledges and sills	Twice a week
	Damp wipe walls and doors	Twice a week
	Damp wipe radiators	Once a week
	Damp wipe water fountains	Once a day
	Dust light fixtures	Every 3 months
	Scrub floors	Every 3 months
	Refinish floors	Every 6 weeks
	Polish metal	Once a month
	Wash partition glass	Once a month
Lavatories	Check soap	3 times a day
	Check towels and tissues	3 times a day
	Damp wipe surfaces	3 times a day
	Empty waste receptacles	Once a day
	Wet mop floor	Once a day

TABLE 4–3 (*continued*)

Area of the Building	Cleaning Jobs	Frequency
	Clean commodes	Twice a day
	Clean urinals	Twice a day
	Clean wash basins	Twice a day
	Damp wipe partitions	Twice a day
	Damp wipe receptacles and dispensers	Twice a day
	Damp wipe ledges and sills	Twice a week
	Damp wipe radiators	Once a week
	Damp wipe walls and doors	Once a day
	Wash walls	Every 4 months
	Wash windows	Once a month
	Dust blinds	Every 2 months
	Wash light fixtures	Once a year
	Wash and rinse floors	Once a day
	Scrub floor	Once a week
Offices	Dust desk tops	Once a day
	Empty and dust ash trays	Once a day
	Dust horizontal surfaces	Once a day
	Dust sills	Once a day
	Sweep floor	Once a day
	Sweep with treated mop	Once a day
	Dust partitions	Once a week
	Dust vertical surfaces	Once a week
	Dust radiators	Once a week
	Damp mop	Once a week
	Carpet sweep	3 times a week
	Vacuum rugs	Twice a week
	Buff floors	Twice a month

TABLE 4–3 (*continued*)

Area of the Building	Cleaning Jobs	Frequency
	Damp wipe doors	Once a day
	Damp wipe water fountains	Once a day
	Damp wipe partitions	Once a day
	Damp wipe sills	Once a week
	Damp wipe radiators	Once a week
	Damp wipe desk tops	Once a week
	Damp wipe door saddles	Twice a month
	Wash partition glass	Every 2 weeks
	Wash partitions	Once a month
	Wash windows	Every 2 months
	Strip floors	Every 6 months
	Mop floors	Every 6 weeks
	Refinish floors	Every 4 months
	Wash walls	Once a year
	Wash furniture	Once a year
	Polish furniture	Once a year
	Dust blinds	Every 3 months
	Wash light fixtures	Once a year
Public Lobbies	Dust radiators	Once a day
	Dust ledges and sills	Once a day
	Dust doors	Once a day
	Dust signs	Once a day
	Clean sand urns	Once a day
	Damp wipe furniture	Once a day
	Damp wipe doors and walls	Once a day
	Damp wipe water fountain	Once a day

TABLE 4–3 (*continued*)

Area of the Building	Cleaning Jobs	Frequency
	Damp wipe ledges and sills	Once a day
	Damp wipe radiators	Once a day
	Wash walls	Every 6 months
	Dust light fixtures	Once a month
	Wash light fixtures	Every 3 months
	Vacuum ceilings	Once a year
	Wash windows	Once a month
	Wash blinds	Every 6 months
	Wash door glass	Once a day
	Polish metal	Every 2 weeks
Stairwells	Damp wipe floors	Once a day
	Damp wipe walls	Once a day
	Damp wipe doors	Once a day
	Damp wipe handrails	Once a day
	Damp wipe ledges and sills	Once a week
	Clean sand urns	Once a day
	Vacuum carpet	Once a day
	Dust blinds	Every 3 months
	Wash blinds	Once a year
	Polish metal	Once a month
	Dust light fixtures	Every 3 months
	Wash light fixtures	Every 4 months
	Dust walls	Every 3 months
	Mop and rinse steps	Once a day
	Scrub steps	Once a week

CONCRETE WALKS AND PAVEMENT

The best preventive maintenance for concrete walks and pavement is to ensure the use of high quality materials and methods during construction. Without this, repairs will be common, and the walks may eventually have to be replaced.

The major maintenance concerns of concrete surfaces fall into three broad categories:

1. The replacement of the surface where it has been cut through to install or repair underground service lines such as water pipes, gas lines, communication cables, and electrical cables
2. The repair of depressions or eroded surfaces caused by excessive wear and resulting from faulty construction
3. Repairs along expansion joints or naturally formed cracks in the surface.

ELEVATORS

A preventive maintenance program used for elevators must not only meet in-house standards but also those standards established by local codes. In fact, in many municipalities and other local areas, all work on elevators must be done by licensed mechanics.

General maintenance schedules as well as recommended types of lubrication and cleaning materials should be provided by the elevator manufacturer. The major reason for scheduling is to establish a set routine for the maintenance mechanic to follow. An effective technique is the use of a checklist that is divided into work that is to be performed during each visit; operations that must be executed on particular calendar dates; and items to be checked and performed semiannually or annually. Table 4-4 provides a checklist that can be used for most elevators on daily, monthly, quarterly, semiannual, and annual inspections.

TABLE 4–4

Maintenance Schedule for Elevators

Elevator Identification: ______

Maintenance Period	Component	Operation
Every Visit		
	Car	Check car flooring for wear.
		Check car grate contacts during operation.
		Lubricate car and counterweight sheave bearings.
		Check condition of switch handle.
		Replace emergency release glass when necessary.
	Hoistway	Check condition and lubrication of rails.
		Check governor and tape tension sheave lubrication.
	Hoistway entrance	Inspect gate cables, body, tracks, and posts.
		Inspect operation of door locks, contacts, checks, and hinges.
		Replace any worn or broken straps on bi-parting doors.
		Check the operation of the car bell system.

TABLE 4–4 (*continued*)

Maintenance Period	Component	Operation
	Machinery, selectors, motor generator, sets, and controllers	Inspect during operation and lubricate as required.
		Empty drip pan.
		Observe brake operation and check the brake lining.
		Adjust brake as necessary.
		Inspect and lubricate machine, automatic contacts, linkage, and gearing.
		Clean rings, brush rigging, commutators, and undercutting as required.
		Clean and lubricate control contacts, connectors, and holders as required.
	Overhead	Lubricate bearings and remove grease from sheaves and bearing boxes.
		Inspect governor parts and lubricate as required.
		Check governor switch.
		Clean and lubricate signal drive mechanism.
	Miscellaneous	Drain air lines.
Once a Month		
	Door and gate	Clean, lubricate, and check brake operation of checks.
		Check keys and setscrews and all contacts.
	Door closure	Clean and lubricate pivot points and sill tips.

TABLE 4–4 (*continued*)

Maintenance Period	Component	Operation
	Selector and commutator	Check brushes, dashpots, traveling cables, chain, brush and pawl magnets, and lubricate as needed.
	Miscellaneous	Check the operation of the signal system in use.
		Inspect drum buffers, rope clamps, slack cable switch, couplings, shafts, and keyways.
Every 3 Months		
	Car	Check the alarm system when in use.
		Clean the lighting fixture.
		Check the emergency switch.
		Check retiring cam device, chain, dashpots, commutator, brushes, cams, pivots, and fastenings.
		Check oil level of AC cam devices.
		Check emergency switch.
		Check adjustments of the car shoes.
		Check safety parts, pivots, setscrews, and switch.
	Leveling	Check leveling switches and operation.
	Machinery	Observe the micro gear case bearing for wear and end play.
		Observe the gear backlash, thrust end play, and bearing wear.

TABLE 4–4 (*continued*)

Maintenance Period	Component	Operation
		Observe worms and gears for end play.
	Motor generator	Clean all commutators.
		Renew or reseat brushes as needed.
	Ropes	Check all fastenings and inspect all ropes for wear and lubrication.
		Check all rope hitches and shackles and equalize the rope tension.
	Tape drive	Clean and lubricate tape drive.
		Clean tapes as required.
	Miscellaneous	Check mainline switch fuses for heating.
		Check compensating chain hitches.
Every 6 Months		
	Bi-Parting doors	Clean chains, tracks, and sheaves, and lubricate as required.
		Check door contacts.
	Car	Check stile channels for bends or cracks.
		Check car frame, cams, and supports.
		Lubricate moving parts of vertical rising or collapsible car gates, and check pivot points, sheaves, guides, and track for wear.
	Car operating box	Check contacts and switches, and clean and lubricate box.

TABLE 4–4 (*continued*)

Maintenance Period	Component	Operation
	Controller	Clean with blower; check the alignment of side weights; lubricate hingepins.
		Check all resistance tubes and grids.
		Check oil in overload relays, settings, and operation.
	Door closer	Check fastenings, operation of checks, and interlocks.
	Door operation	Check shaft bridges, tapered pins, alignment, operation of cams, and rollers.
	Machinery	Seal all small leaks.
	Pit	Lubricate compensating sheaves and check hitches.
		Check oil level in oil buffers.
		Check governor and tape tension sheave fastenings.
	Selector	Check traveling nut and gears for wear.
	Tape drive	Check hitches and broken tape switch.
	Miscellaneous	Lubricate indicator dials and pulleys.
		Clean car grillwork and stile channels visible from inside of the car.
Once a Year		
	Car	Check car enclosure steadying plates.
		Check clearances for car safety shoes.

TABLE 4–4 (*continued*)

Maintenance Period	Component	Operation
	Controller	Clean and check fuses and holders.
		Check controller connections.
	Guide shoes	Lubricate stems and adjust as necessary.
	Hall	Check hall button contacts, springs, and wiring.
		Clean and lubricate as required.
	Hoistway	Clean and check guide rails, cams, fastenings, and counterweights.
		Inspect limit and terminal stop switches.
		Lubricate pins and rollers.
	Hoistway doors	Clean and lubricate tracks and hangers.
		Check upthrust and adjust as required.
		Fill and adjust checks.
		Check bottom gibs, struts, sills, headers, fastenings, and adjust door controls.
	Machinery	Remove, clean, and lubricate brake cores on DC brakes.
		Clean linings, if required, and check for wear.
	Motor generator exciters	Clean armatures and motors with the blower of a vacuum.
		Check armature and rotor clearances.
		Check motor connections.
		Change oil in bearings.

TABLE 4–4 (*continued*)

Maintenance Period	Component	Operation
	Rail guide	Lubricate wheel bearings.
	Sheaves	Adjust sheaves so that they are tight on shafts.
		Check sound spokes and rim for cracks.
	Traveling cable	Check for wear, insulation, and hangers.
		Check upthrust and adjust as required.

COMMON PROBLEMS

Regularly scheduled maintenance inspections and jobs will not only extend the serviceability of elevators but will also uncover problems. Some of the common problems are discussed briefly here.

Brake Problems. Brake problems include inability to hold 125 percent of a full load, noisy lifting under load, worn lining, frozen pins, rough or dirty pulleys, excessive brake shoe clearance, misaligned brake shoes, and improperly set brake switches.

Car Problems. Within the car itself inspection will expose worn flooring, a missing or improperly secured emergency exit, broken buttons, broken phone, nonlocking car switch handles, nonworking lights or alarm or stop buttons, broken formica, plugged gate contacts, dragging gates, and worn gate guides.

Bottom of the car inspection will expose worn safety shoes, frozen drums, insufficient turns on the safety drum, crossed ropes, loose or missing safety rollers, worn gibs, unsafe car shoes, and loose lead sheaves.

Top of the car inspection will expose loose fastenings and plates, rope vibration, worn gibs, lost lay in ropes, nonfunctioning switches, unaligned door operators, and oil leakage.

Controller Problems. Problems with the elevator controllers include wrong fuse capacity, broken or loose leads and connections, broken resistance, poor or worn contacts, improper contact spring tension, worn pins and bushings, sticking switches, overheated coils, and open or short circuiting.

Hall Problems. Hall problems include broken buttons and emergency key glasses, missing keys, inadequate lighting, and missing or nonworking fire extinguishers.

Hoistway Problems. Worn cables, worn hoist and governor ropes, defective doors, wear in door pins and chains, wear of bushings and shoes, loose and worn tracks, dirty and sticking counterweight piston, unaligned rails, loose rail fastener and brackets, and worn interlock fingers and springs can all affect hoistway performance.

Machine Problems. Broken or loose gear rim bolts, excessive backlash between worm and gear, cracked sheave spokes, loss of gear lubrication, and uneven sheave groove are common elevator machine problems.

Motor and Generator Problems. Oily armatures or stators, poor brush-to-spring pressure, worn brushes, flat commutator spots, pitted bars, bearing wear, and lost ground clamps are common motor and/or generator problems.

Pit Problems. The elevator pit may have an improperly lubricated car buffer, a sticking piston, worn bearings, stretched rope, and switches out of adjustment. Debris may also be present in the pit.

FLOORS AND WALLS

All good building maintenance programs include good floor care. The exact type of floor maintenance activities required depends on the type of flooring material, the amount of traffic, the type of activities conducted on the floor, atmospheric conditions, and the appearance level desired. Table 4-5 lists the major types of flooring material and recommended maintenance procedures for each.

Although most maintenance mechanics will not be responsible for painting large areas, the condition of painted surfaces may be a concern in the overall maintenance of a building or facility. Table 4-6 lists the common causes of paint deterioration and will help the mechanic diagnose a paint problem and, if necessary, arrange for its correction.

TABLE 4–5
Floors and Floor Maintenance

Type of Flooring	Description	Maintenance Considerations
Asphalt tile	Mix of fibers, pigments, and inert fillers, bound with asphalt or resin binder. Deteriorates under exposure to oils, greases, and solvents such as gasoline, naphtha, turpentine, kerosene, and carbon tetrachloride.	Use a good quality mild soap, neutral cleaner, or synthetic detergent. Do not use alkaline or caustic cleaners.
Concrete floors	Sand and gravel bound with cement	Heavy traffic area floors should be sealed with a modified phenolic or epoxy resin floor sealer. Clean sealed floors with a neutral, synthetic, free rinse detergent. When exposed to greases, oils, and mild acids and alkalis, use a chlorinated rubber base seal.
Cork tile	Ground cork bark molded and compressed.	Clean with neutral synthetic detergent and avoid all types of alkali cleaners. Never expose to naphtha, gasoline, or similar solvents, oily rags and mops, or sprays.

TABLE 4–5 (*continued*)

Type of Flooring	Description	Maintenance Considerations
Linoleum or linoleum tile	Composed of oxidized linseed oil, fossil, resins, or an oleo-resinous binder with ground cork, wood flour, mineral fillers, and pigments.	Use solvent or water-type wax or emulsified resin finish. Use neutral stripper or a nonalkali cleaner. Deteriorates when exposed to any alkali and is stained by grease and oil.
Marble floors		Use neutral cleaner and damp mop daily. Pre-wetting helps to prevent salts from entering the pores of the floor. May be stained by oily sweeping and frequent use of ammonia. Never use abrasive cleaners. Acid and strong alkali cleaners are not recommended and will leave a dull finish.
Rubber tile	A mixture of rubber (natural, synthetic, and/or reclaimed), inert fillers, and pigments.	Synthetic detergents are approved with this tile. Mild alkaline cleaners (except TSP) can also be used occasionally. Avoid sweeping compounds, and substances containing gasoline, kerosene, naphtha, benzene, turpentine, mineral solvents, or harsh alkalis.

TABLE 4–5 (*continued*)

Type of Flooring	Description	Maintenance Considerations
Terrazzo floors	Mix of marble chips and concrete. Can be damaged by both acid and alkali cleaners.	Should be sealed. Use only a neutral cleaner.
Vinyl tiles	May be vinyl asbestos; homogeneous or vinyl plastic; calendered vinyl mix or backed vinyl; or rotogravure printed or vinyl coated gravure.	Use water emulsion wax or spirit wax. Can also take solvent type cleaners, waxes, and sweeping compounds.
Wood floors		Should be filled and sealed with a water-resistant finish. Damage and deterioration will occur when using harsh or strong alkali cleaners.

TABLE 4–6
Causes of Paint Deterioration

Type of Deterioration	Causes
Alligatoring—cracks that appear as an alligator's hide and which may or may not reach the base surface	Use of a soft coat over a hard coat of paint.
Blistering—blisters that develop on the surface and usually break open	Moisture in the base surface that pushes up to the skin. Drying occurs too rapidly.

TABLE 4–6 (*continued*)

Type of Deterioration	Causes
Bubbling—bubbles that appear on the surface, usually found on wood.	Moisture or sap rising to the surface to form bubbles.
Chalking—premature chalking or fine powdering appearing on the surface	Poorly formulated paint, or paint applied over a porous and extremely weathered base.
Checking—the cracking and opening of paint film in spots	Shrinkage of paint film due to uneven coating and poor bonding qualities of the paint. Improper compounding resulting in the paint ingredients drying in improper sequence. Excessive or too little solvent in the paint.
Chipping or flaking—the breaking away of paint film	Expansion or shrinkage of base at different rates from the surface coat. Extreme changes in temperature during paint application. Improperly prepared surface.
Discoloration—various sections of off shade or dark spots	Impurities under the paint surface. Attack by chemical gases or fumes.
Peeling—paint film either peeling or chipping away in large sections	Moisture beneath the base coat. Breaking or falling of caulking compounds.
Rusting—streaked surfaces	Surface improperly cleaned.

TABLE 4–6 (*continued*)

Type of Deterioration	Causes
Spalling—cracking, splitting, or breaking off of paint from brick or concrete	Excessive moisture within the brick or concrete.
Wrinkling—a grainy appearance on a surface	Applying too thick a coat of paint at one time, causing improper drying. Paint applied in unseasonable weather.

LIGHTING

The maintenance of lighting facilities usually involves replacing lamps and cleaning fixtures. The specific program selected by the mechanic depends on the cost of maintaining an area versus the requirements. In many cases it is necessary to review possible energy conservation measures, such as the use of high intensity discharge lighting, as a way of improving efficiency.

REPLACING LAMPS

Spot Replacements. The replacing of burned out lamps as they fail is known as spot replacement. The major disadvantage of this technique is the tedious and inefficient process of locating burned out lamps.

Group Relamping. Group relamping is the replacement of lamps before they burn out. This technique is generally considered more advantageous than spot replacement. It results in better lighting levels and can be accomplished at any convenient time—for example, during shutdowns or after work hours. This approach also reduces the damage and burning out of starters and ballasts caused by blinking and slow starting aged lamps.

CLEANING FIXTURES

Cleaning lighting fixtures is important, though often overlooked. An accumulation of dirt and dust can reduce output as much as 50 percent.

METALS

Since much machinery and other equipment is metal or has metal components, it is important for the maintenance mechanic to understand the major problem associated with metal—namely, corrosion. Corrosion is a destructive attack on metal that may be caused by either chemical or electrochemical actions. Although some metals are resistant to some forms of corrosion—for example, stainless steel (chromium added to steel) is resistant to rusting in the atmosphere—corrosion is still a common problem with many metals. Table 4-7 lists the common types of corrosion and their causes.

TABLE 4–7
Corrosion of Metals

Type of Corrosion	Causes
Uniform corrosion—thinning and loss of material	Exposure to attacking environmental conditions.
Pitting—highly localized with slight uniform corrosion	Most common with aluminum and stainless steels exposed to moisture with metal chlorides present.
Stress corrosion cracking, also known as season cracking and caustic embrittlement—the failure of the metal under load or stress	Occurs in most metals and alloys, with each one having its own cause. Cold formed or rolled brass exposed to ammonia.

TABLE 4–7 (*continued*)

Type of Corrosion	Causes
	Austenitic stainless alloys exposed to chlorides. Monel exposed to hydrofluosilic acid. Steels exposed to caustic solutions.
Hydrogen embrittlement—microcracking, blistering, and loss of ductility	Irons and steels exposed to water solutions that contain hydrogen sulfide.
Crevice corrosion—localized corrosion on surfaces partially protected from contact with corroding solutions	Cracks or openings in protective finishes or membranes.
Galvanic corrosion—electrolytic attack on metals	Contact between two metals, such as zinc and steel exposed to water.
Graphitization—formation of a spongy corroded mass	Reaction between graphite and iron constituents found in cast iron.
Fretting corrosion—localized corrosion	The rubbing of two surfaces that removes the protective surface coating.
Oxidation—metal scaling	Usually caused by high temperatures. The formation of metal oxides.
Sulfidation—intergranular attack	Metals exposed to sulfur-bearing atmospheres at elevated temperatures. (Nickel and its alloys are not suitable for sulfur atmospheres.)

ROOFS

Roofs, like any other component of a plant, require regular upkeep and care. When inspecting roofing, do a complete survey of the entire building and its use, for the roofing system is an integral part of the entire structure. Table 4-8 provides a checklist that should be used when making a roof inspection.

TABLE 4–8
Roof Inspection

Roofing Component	Potential Defects
Walls	Signs of building movement, such as cracks and joint separation.
Interior bearing partitions	Signs of movement and cracking.
Lintels across doors and windows	Openings, cracks, or gaps.
Interior and exterior drains	Dampness or flow of water.
Underside of roof deck	Dampness, deterioration, and rot.
Copings	Deterioration of joints and cracking.
Sheet metal components	Metal corrosion or deterioration and loosened fastenings.
Composition of flashing	Damage, material deterioration, cracks, breaks, and punctures.
Composition edging	Material deterioration, breaks, cracks, and punctures.
Main roofing system	Debris laying about, inappropriate storage of materials, broken or damaged drainage sumps, physical damage, and material or surface deterioration.

Three techniques are used to maintain roofs: spot repair, recoating, and recovering.

1. *Spot repair.* Spot repair is patching and any other repair of localized damage.
2. *Recoating.* Recoating is the application of waterproofing material for protection. It should only be done if the roof coating is in generally good condition but has a few thinning areas.
3. *Recovering.* Recovering is a major job that leaves the old roof system intact while covering it with a completely new roofing system. Since all old roofs should be assumed to have a high moisture content, recovering is not recommended. The old roofing system should be removed prior to the installation of a new roof.

5
Electrical Equipment

Maintenance and machine mechanics are not trained as electricians, nor are they expected to perform jobs usually handled by such tradespeople. However, they are required to have basic knowledge and skills in electrical work, for there are many times when electrical equipment must be maintained by the mechanic. This chapter discusses preventive maintenance and troubleshooting procedures for electrical equipment frequently encountered by the maintenance and machine mechanic as integral components of either building and plant facilities or of machinery.

BATTERIES

Batteries are devices that transform chemical energy into electrical energy. In any commercial setting, it is important to select the correct battery. To do that the mechanic must know the applications and service life requirements of different types of batteries. It is also important to know the basic causes of and remedies for common battery problems (see Table 5-1).

TABLE 5–1
Troubleshooting Battery Problems

Problem	Probable Causes	Recommended Corrections
Battery will not take a charge	Blown or missing DC charging circuit fuse	Replace fuse.
	Loose cable connection or circuit in charging receptacle or open plug	Tighten connections or close circuit.
	Open AC line switch	Close line switch.
	Circuit open or in control lead	
	Too low a charging rate	Check accuracy of ammeter.
	Charging plug not fully inserted in receptable	Insert plug.
	No voltage supply	Check field circuit and correct if it is open. Check brush contact to armature and replace or adjust.
	Bus voltage too low	Check tap setting in rectifier. Too low of a voltage output from the power source.
	Reversed connections to the charging receptacles	Correct connections.

TABLE 5–1 (*continued*)

Problem	Probable Causes	Recommended Corrections
Battery takes too long to charge	Poor connections in charging circuit	Check connections, lugs, boltings, charging leads, plugs, and receptacles.
	Overdischarging of battery	Check and correct.
	Insufficient starting rate in charging equipment	Open control circuit and correct.
	Voltage relay connected for less cells than are in the battery	Provide for additional cells.
	AC voltage too low	Install line with a greater capacity.
	Primary transformer taps set for incorrect voltage	Correct voltage settings.
	Temperature of charging control equipment higher than battery operating temperature	Lower area temperature by improving ventilation or relocate to a cooler area.
	Reversed charging leads	Check and correct.
	Battery not in correct charging circuit	Check and correct.
	The charge not stopped when the battery is fully charged	Check and correct.
Battery will not work for rated period of time	Specific gravity and cell voltages uneven	Apply equalizing charge.
	Low electrolyte level	Fill cells with electrolyte.
	Battery not charged or not fully charged before pressed into service	Charge battery.
	Cell leakage	Replace broken jars.

TABLE 5–1 (*continued*)

Problem	Probable Causes	Recommended Corrections
	Cells cutting out of battery	Check and correct.
	Incorrect battery size and rating placed into service	Check and replace.
	Specific gravity below normal	Check and correct.
	Impurities in electrolyte	Drain and clean cells, replace contaminated electrolyte.
	Operator overloading battery	Check and correct.
	Series field in motor shorted or grounded	Clear grounds and provide proper insulation for wiring.
	Damaged armature	Repair.
Overheating of battery while on charge	Too high finish rate	Check and correct.
	Too long of a high charge rate	Reduce voltage operating point of the voltage relay.
	Charging time incorrectly set or malfunctioning	Check and correct.
	Charge not stopped by automatic device	Check mechanism, check for open control leads, check function of voltage relay, make sure voltage relay is connected for correct number of cells, and check meter for correct calibrations.
	Poor ventilation	Check and correct.
	Internal shunt	Check and repair or replace.
	Separators worn through	Check and repair.

TABLE 5–1 (*continued*)

Problem	Probable Causes	Recommended Corrections
Overheating during battery discharge	Overdischarge	Check and correct.
	Undercapacity rated battery for service	Check and correct.
	Poor ventilation	Check and correct.
	Excessive load	Check and correct.
	Battery not fully charged prior to service	Fully charge battery.
	Low electrolyte level	Fill cells with electrolyte.
	Atmospheric temperature too high	Check and provide for better temperature insulation or move to cooler area.
Electrolyte level low	Broken or cracked jar	Repair or replace.
	Neglected water level checks and maintenance	Maintain schedules.
	Excessive overcharging	Check voltage relay, timer, and change rate relay.
Unequal cell voltage	Overdischarge	Equalize charge.
	Internal shunt	Check and correct.
	Excessive dirt buildup on the top of battery	Wash top of battery.
	Low electrolyte level in cells	Check and add fluid.
	Impurities in the cell	Drain cells and add pure electrolyte.
	(Also see following problem.)	

TABLE 5–1 (*continued*)

Problem	Probable Causes	Recommended Corrections
Unequal specific gravity between cells	All problems listed under unequal cell voltage	
	Too much water added to cell	Correct cell fluid levels.
	Cracked jar	Repair or replace.
	Vent caps loose or missing during service	Check and correct.
	Leak in sealing compound	Check and correct.
	Broken cover	Check and replace.
	Material in the cell neutralized	Remove and replace with fresh electrolyte.

LEAD-ACID BATTERIES

Perhaps the most common type of battery used today is the lead-acid storage battery, which is found in a wide variety of equipment, vehicles, and machinery. These batteries are classified as lead-acid because the electrolyte (fluid) used is an acid and the plates are primarily lead.

Batteries used in most commercial and industrial equipment are used in cyclical operations. In other words, the battery will be either in or out of use, and when not in use, will be in the process of being charged. Hence, one of the most important maintenance procedures required is that of battery charging. (See Chapter 8.)

Similar to all other maintenance activities, battery condition is determined by maintenance records, inspection and testing, and routine care. Table 5-2 describes the procedures recommended for the routine maintenance of lead-acid storage batteries.

TABLE 5–2
Maintenance of Lead-Acid Batteries

Maintenance Activities	Description
Recordkeeping	Include in the daily records the date, battery number, equipment battery was taken from, specific gravity of battery when placed in charge, temperature of the pilot cell, time put into and taken off charge, specific gravity when taken off charge, and the piece of equipment battery is placed on.
Test discharge	Discharge at a standard rate (for example 6 hour rate) given by battery manufacturer; record individual cell and overall battery voltage periodically, and measure specific gravity after testing for uniformity.
Internal inspection	Conduct if test discharge is less than 80 percent of rated capacity. Check for plate wear, grid integrity, and cell contamination.
Routine care	Inspect weekly to make sure all connections are tight and remove dirt or dust. Wash with water, neutralize any acid accumulation with ammonia or baking soda once a month. Keep terminals and other metallic parts free from corrosion. Check level of electrolyte (fluid level), and only use water approved for battery use. Be sure vent plugs are tightly in place.

NICKEL-CADMIUM-ALKALI STORAGE BATTERIES

Nickel-cadmium-alkali, or Nicad, batteries have only recently been adopted in the United States, though they have been employed in a variety of situations in Europe for many years. The positive material is nickelic hydroxide, which is mixed with graphite; the negative material is cadmium oxide, and the electrolyte is potassium hydroxide (caustic potash).

Nicad batteries possess a good charge acceptance, function well at high discharge rates, and outperform other common storage batteries at low temperatures. Not designed for cyclical applications, they are commonly used for engine starting, emergency lighting, communications, alarms, and switch gear.

The primary maintenance concerns with Nicad storage batteries are maintaining electrolyte level and specific gravity and keeping the exterior casing clean. Table 5-3 lists recommended maintenance procedures.

TABLE 5–3
Maintenance of Nickel-Cadmium-Alkali Batteries

Maintenance Procedure	Description/Comment
Maintain proper electrolyte level, gravity, and purity.	Regularly check electrolyte level; add water to maintain solution level and renew electrolyte as required.
Guard against stray currents and shorts.	Make sure metal cells do not come in contact; keep cases and covers clean; never overfill cells; dress cables away from cell tops; and during battery voltage checks, look for any voltages between any terminals and ground.
Maintain tight connections.	Make sure that there is good electrical contact at all terminals and that seals around posts and vents are tight.

TABLE 5–3 (*continued*)

Maintenance Procedure	Description/Comment
Avoid sulfuric acid.	Be sure that no sulfuric acid is introduced, as it will ruin any nickel-cadmium-alkali battery.
Avoid high temperatures.	Make sure that batteries are not regularly exposed to temperatures exceeding 100°F.
Keep away from all open flames and sparks.	Remember that hydrogen and oxygen released by the battery are highly explosive.
Repaint pocket cells.	Paint pocket cells and trays as needed.

BRAKES

There are two general types of electric brakes: direct and alternating current. Direct current brakes may be operated by a series of coils or shunt coils; alternating current brakes operate only by shunt coils. All brakes, however, are classified by the type of operator used: alternating current solenoid operator, direct current solenoid operator, alternating and direct current brake operator with clapper-type magnet, elec-

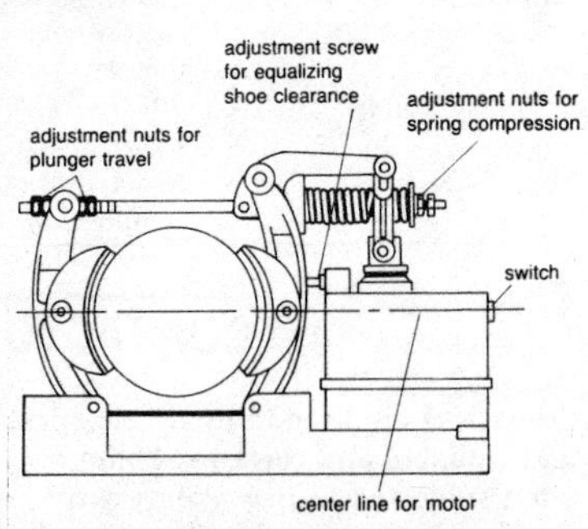

Figure 5-1. Brake adjustments usually involve plunger travel, shoe clearance, or spring compression.

trohydraulic brake operator, or disc-type brake operator.

The maintenance mechanic must ensure that the parts of brakes are properly maintained and adjusted. Table 5-4 lists brake parts and related maintenance requirements.

TABLE 5–4
Maintenance of Brake Parts

Parts	Maintenance Requirements
Shoe-type brake mounting	Check vertical and horizontal center line against manufacturer's outline drawings.
Brake wheels	Check for damage, cracking, brakes, scores, or rubbing. Replace brake wheels if they are not balanced. Brake wheels should never exceed operating limits nor be used at temperatures exceeding 392°F.
Brake linings	Check linings periodically for wear and replace when needed. Clearance on the brake shoes should be equalized.
Bearings and pins	Lubricate with small amounts of oil or light grease. Do not over lubricate.
Manual release devices	Check reset features and interlock manual release.
Torque adjustments	Check braking torque on the shoe brake so that it is proportioned to the spring compression. Do not adjust the spring so that the operator cannot open the brake under all operating situations.
Brake adjustments	Readjust brakes as needed for plunger travel, shoe clearance, and spring compression. (See Figure 5-1.)

CONTACTORS

Contactors are used in many electrical devices to handle low and high voltages and currents. Commonly used in industry, they are often subjected to heavy duty work, especially in milling operations, and so should be of sturdy construction. When main contacts interrupt motor currents and voltages, their current ratings may be high. Contactors are most frequently made of copper, although silver and alloy contactors are found in many devices.

Maintenance programs should begin as soon as contactors are placed into service. Before inspection begins, all power should be turned off from the starter, and fuses should be removed. Recommended inspection procedures are listed in Table 5-5.

Contactor problems occur from time to time. Table 5-6 lists the most common problems, their causes, and possible remedies.

TABLE 5–5
Inspection of Heavy-Duty Contactors

Inspection Checks	Points to Observe
General inspection	Look for loose, missing, broken, or corroded components. Check pivot pins, cotter pins, springs, and other parts.
	Do not lubricate moving parts.
Arc boxes	Check for broken or eroded parts and grid plates. Make sure that there is no excessive collection of contact material on the surface of insulating parts.
	Replace broken or eroded parts.
	Clean parts with excessive material buildup.
Blowout coils	Check for discoloration, shorted turns, arcing, and loose hardware.
Contacts	Clean contacts of oxides or beads of contact materials by sanding or filing.
	Do not file or sand silver or alloy contacts unless buildup material is excessive.
	Replace contacts with correctly rated replacements and realign and set overtravel.
Electrical operation	Observe magnet operation to be sure it opens and closes cleanly.
	Check that the armature is fully sealed in the closed position during operation.
	Check sample of oil for proper level and type.
	Operate contactor under load and check for any abnormalities such as arcing.

TABLE 5–5 (*continued*)

Inspection Checks	Points to Observe
Insulators	Clean dust and dirt from parts.
	Replace defective parts as indicated by carbonized tracks or cracked or broken insulators.
Magnet assembly	Check for dirt, corroded pole faces, pivot points, and other moving parts.
	Check for loose, broken, or missing shading coils, residual shims, and hardware.
	Inspect operating coil for any damage.
	Inspect armature for mechanical interference and friction.
Shunts	Replace broken or frayed shunts.
	Clean connection points as needed.
Terminals, contact supports, bus bars, and connectors	Look for any discoloration, which will be an indicator of overheating.
	Clean discolored points and tighten hardware.

TABLE 5–6

Troubleshooting Contactor Problems

Problem	Probable Causes	Recommended Corrections
Contactor overheating	Too high a load current or loose connections	Reduce load or use larger contactor. Clean discolored or dirty connections and secure tightly.

TABLE 5–6 (*continued*)

Problem	Probable Causes	Recommended Corrections
	Contact force too low and/or contact overtravel	Replace contacts and springs. Adjust overtravel.
	Accumulation of oxides or other material on face of contact	Clean with fine abrasive or file.
	Load applied to contact in excess of rating	Check and correct. Replace with proper contact.
	High ambient temperature	Reduce load; improve ventilation; use larger contact, and/or locate contact in cooler area.
	Too small line and/or load cables	Check and replace with properly rated cable.
Contacts welded together	Overtravel or contact force too low	Replace contacts and springs, adjust overtravel.
	Magnet stalling or hesitating at points of contact	Replace.
	Magnet chatter	Check overtravel and contact force; replace control device or move to location that has less chatter.
	Contacts bounce when closing	Check and adjust coil overvoltage condition.
	Contacts touch when opening	Check and repair.
	Contacts misaligned	Check and adjust.

TABLE 5–6 (*continued*)

Problem	Probable Causes	Recommended Corrections
	Severe jogging duty	Use larger contactor or replace with weld-resistant contacts. Reduce rate of jogging cycle.
	Too high inrush current	Check and use larger contacts or replace with weld-resistant contacts. Adjust acceleration time and/or operating sequence.
	Starter mounting vibration	Improve starter support, shield from vibrations, or move to area with less vibration.
Service life of contact less than rating	Poor contact force	Adjust overtravel. Replace contacts and springs.
	Contact bounces upon opening and/or closing	Check and correct coil overvoltage. Replace defective components.
	Magnet chatters	Check overtravel and contact force; replace control device, or move to area with less chatter.
	Accumulation of abrasive dust on contacts	Clean with non-abrasive material. Enclose in dust protective device.
	Jogging cycle too harsh	Use larger contact. Replace with more appropriately rated contact. Reduce jogging cycle rate.
	Wrong type of contactor used	Replace oil immersed contactor with air break contactors.

TABLE 5–6 (*continued*)

Problem	Probable Causes	Recommended Corrections
Poor arc interruption	Arc box not in use or damaged	Check and install or repair.
	Soiled arc horns or steel grid plates	Remove dirt, paint, and any other insulating material.
	Short-circuited or reversed blowout coil	Check and correct. Replace blowout coil.
	Low oil level or worn-out oil in oil immersed contactor	Replenish tank to proper level with new oil.

CONTROLS

Electrical controls are used to regulate electrical flow in the operation of machinery and equipment. Examples of common controls are motor starters, contactors, switches, and transformers. To ensure the success of a maintenance program for any or all of these devices—many of which are discussed separately—it is important at the outset that they be of good quality and be correctly installed.

Although most electrical equipment is designed to operate under a wide range of conditions, it can also be damaged easily by inappropriate working conditions. Moisture, extreme temperatures, corrosive atmospheres, dust, and other conditions can significantly affect the performance and service life of an electrical apparatus. Generally, a three-point practice for maintaining electrical equipment should be:

- Keep it clean.
- Keep it dry.
- Keep it tight and friction-free.

Maintaining Electric Controls

1. Test all initial installations and make sure that they perform satisfactorily before accepting them.
2. Be sure that all electrical devices are installed in such a way and location that they are easily accessible for inspection and repair.
3. Be sure that enclosures in which devices are housed are selected based on operating conditions—for example, dusty atmosphere, wet locations, hazardous locations, or corrosive atmospheres.
4. Be sure that an adequate supply of correct renewal parts is kept on hand.
5. Keep control centers clean and dry.

ELECTRONIC CONTROLS

Most electronic controls used today are designated to operate for long periods of time without significant maintenance. In past years, electronic tubes were used extensively for industrial controls. However, since the 1960s, tubes have been replaced by transistors, diodes, and other similar electronic components. In most situations, these devices are inspected and tested by licensed electricians or electronic technicians.

LIGHTING

The primary maintenance requirement for lighting systems is cleaning fixtures, which was discussed in the preceding chapter. In addition to cleaning and lamp replacement, the maintenance mechanic should also be able to analyze the lighting system for its efficiency and maintenance requirements. Table 5-7 summarizes different checking and inspection procedures.

TABLE 5–7
Analyzing Existing Lighting Systems

Maintenance Factors	Description of Methods
Lighting intensity	Measure lighting intensity for each individual area and lighting system. Make use of a portable photoelectric light meter positioned at various stations.
Lamp outages	Record and count all lamp outages, regardless of reasons.
Lamp depreciation	Determine anticipated life expectancy of lamps and compare to actual results. Calculate average lamp life for use in replacement program.
Socket voltages	Measure voltage at lamp sockets or ballast terminal leads and record findings.
Luminaire inefficiency	Obtain the light distribution characteristics and lighting efficiency from manufacturer. Compare these with units in use.
Room-surface reflectance	Determine the reflection factors for ceiling, walls, floors, furniture, and other objects.
Dirt and dust	Take footcandle readings before and after cleaning and compare.
Maintaining acceptable lighting intensity	Based upon the preceding factors, identify problems and maintain or adjust existing lighting intensities.

MOTORS

Electric motors are used to convert electrical energy into mechanical energy. They are available in sizes ranging from fractions of a horsepower to several thousand horsepower, and come in a variety of speeds. Many commercial motors are self-starting and reversible. Though there are many different types of motors, all electric motors fall into one of two groups: alternating current (AC) and direct current (DC).

Alternating current motors are the most common because of the easy availability of AC power. The most common type of AC motor is the *induction motor,* in which current is induced in a rotor as its conductors cut lines of magnetic flux. When constant speeds are needed, a *synchronous motor* is employed. Situations requiring either AC or DC power dictate the use of the *series motor.* When a high starting torque is required, the *universal motor* is recommended.

Direct current motors are characterized by ease of speed control and, in the series connected motor, by the ability to produce large torque under load without the need for excessive current.

Although there are design differences between AC and DC motors, there are several maintenance practices that should be followed with each. These are described in Table 5-8.

TABLE 5–8
Maintenance of Electric Motors

Maintenance Practice	Comment
When not in use, keep the motor off line.	This practice will save wear of brushes, commutator, and bearings.
Unless the motor has been so designed, do not leave the field circuit excited.	Temperature of the shunt fields should be taken and should not exceed 194°F. Situations requiring field excitement dictate that all maintenance personnel be sure to open field circuit before working on the motor.
Keep the motor clean.	Motors should be clear of metal dust or cuttings that can be drawn into the windings and pole pieces, since magnetic attraction will draw iron and steel parts into the air gap and can damage the windings.

TABLE 5–8 (*continued*)

Maintenance Practice	Comment
Reassemble the motor.	Proper air gaps must be maintained and checked by inspecting the bore of pole faces prior to disassembly. When reassembling, be sure that the poles and liners are aligned to original positions.
Check for worn parts and those frequently replaced.	This will help to determine anticipated repairs. Always have a supply of replacement parts.
Check and lubricate bearings.	Handle bearings carefully; remove and install properly; and lubricate with proper lubricant.

Table 5-9 lists problems that commonly occur with motors, their probable causes and solutions, and then deals with problems specific to AC motors and DC motors.

TABLE 5–9
Troubleshooting AC and DC Motor Problems

Problem	Probable Causes	Recommended Corrections
General Motor Problems		
Overheated bearings (general)	Shaft bent or sprung	Check and replace or straighten.
	Belt pull overload	Decrease belt tension.
	Pulley too great a distance from bearing	Move motor closer to the bearing.
	Too small a pulley diameter	Replace pulley with larger pulley.
	Out of line drive	Realign drive.

TABLE 5–9 (*continued*)

Problem	Probable Causes	Recommended Corrections
Overheated sleeve bearings	Clogged or obstructed oil groove	Clean oil grooves and bearing housing and lubricate.
	Oil rings damaged or bent	Replace or repair.
	Too heavy or too light an oil weight	Check and use recommended oil.
	Poor lubrication	Check reservoir and fill to proper level.
	Excess end thrust	Reduce thrust or use external methods to handle thrust.
	Worn bearings	Replace.
Overheated ball bearings	Poor lubrication	Check and make sure there is sufficient grease in bearing.
	Grease deterioration or contaminated lubricant	Remove lubrication, clean bearings, and add new grease.
	External heat source	Improve ventilation or remove heat source.
	Bearing overload	Check and adjust alignment, side and end thrusts.
	Rough brakes and/or broken ball	Replace and clean housing.
Oil leakage from overflow plugs	Unsecure or loose plug cover	Tighten or add proper gasket.
	Damaged overflow plug	Replace.

TABLE 5–9 (*continued*)

Problem	Probable Causes	Recommended Corrections
Dirty motor	Poor ventilation	Dismantle motor and clean thoroughly.
	Clogged rotor winding	Clean grind and undercut commutator. Windings should be cleaned and finished with insulating finish.
	Inside coating of bearing and bracket	Clean and wash with solvent.
Wet motor	Exposed to dripping, drenching, or other wet conditions	Relocate motor, protect with cover or casing, dismantle, clean, and dry parts.
DC Motor Problems		
Motor fails to start	Open circuit	Check switch circuit and power supply.
	Bushes not in contact with commutator	Check bush springs and bushes, and replace if not functioning or worn.
	Motor or main drive bearings frozen and locking armature	Replace or recondition bearings.
Motor starts, stops, and then changes direction of rotation	Reverse polarity	Check generating unit and wiring for cause of reverse polarity.
	Shunt and series fields fighting or "bucking" each other	Correct polarity.
Motor does not reach full speed	Overloading of motor	Reduce load or friction. Check bearings to see that they are properly lubricated.

TABLE 5–9 (*continued*)

Problem	Probable Causes	Recommended Corrections
	Low voltage supply	Check incoming voltage and compare to motor rating.
	Low starting resistance	Check condition of starter.
	Short circuiting in windings or between bars	Inspect armature and windings for signs of burning.
	Motor not in neutral	Check to see that motor is in true neutral setting.
	Cold motor	Increase motor load and add rheostat for speed setting.
Motor running too fast	Too much voltage for motor rating	Check and correct voltage.
	Light load	Increase load.
	Shorted shunt field coil	Check and replace shunt coil.
	Reversed shunt coil	Check and reconnect leads correctly.
	Shorted series field coil	Check and reconnect leads correctly.
	The neutral setting off neutral	Test and reset neutral.
	Poor motor ventilation resulting in hot shunt field	Check ventilation and hot field causes and correct.
	Unstable speed regulation	Inspect motor and check series field.
	Excessively strong commutating pole, or commutating pole air gap not large enough	Check and correct coils or air gap.

TABLE 5–9 (*continued*)

Problem	Probable Causes	Recommended Corrections
Motor running too slow	Too low voltage for motor rating	Increase voltage supply.
	Overload on motor	Check bearings and friction for signs of overloading and correct.
	Cold operating motor	Load is too light for motor rating; switch to a smaller motor.
	Change in neutral setting	Check and adjust.
	Shorted coils or commutator bars	Remove and repair armature.
Motor overheats and/ or runs hot	Overloading, drawing less current than rating	Measure power line and correct.
	Voltage overrated	Reduce voltage to motor rating.
	Poor ventilation	Remove restrictive condition.
	Short in coil causes drawing of too much current	Repair or replace armature coil.
	Off-center rotor causing friction heat and use of excessive current	Check condition of bearing and correct. Realign rotor.
Hot armature	Shorted punchings and high iron loss	If full slot metal wedges were used for balancing, replace.
	Uninsulated punchings	Replace core and rewind armature; check core temperature and be sure it does not exceed 167°F.

TABLE 5–9 (*continued*)

Problem	Probable Causes	Recommended Corrections
Commutator running hot	Tension of brushes too high	Reduce pressure.
	Brushes not on neutral	Reset neutral.
	Too abrasive a grade of brushes	Check and replace with recommended brushes.
	Bars shorted	Check and correct.
	Hot core and coils	Check and correct.
	Poor ventilation	Correct restrictive condition.
Hot fields	Poor ventilation	Check and remove restrictive condition.
	Too high voltage	Correct voltage to motor rating.
	Short in turns or ground turns	Check, repair or replace turns.
	Different coil resistances	Correct to equal resistance of each coil.
Motor vibration	Out of balance armature	Remove armature and balance.
	Misalignment	Check and realign.
	Pulley loose or eccentric	Tighten or balance pulley.
	Whipping action in the belt or chain	Adjust tension.
	Poor gear and pinion mesh	Repair or replace, and realign.
	Bent shaft	Replace or repair.
	Poor motor foundation or mounting	Reinforce mounting.
	Coil too small	Replace with proper size coils.

TABLE 5–9 (*continued*)

Problem	Probable Causes	Recommended Corrections
Sparking at brushes, or does not commutate	Not set in neutral	Check true neutral and correct.
	Rough or eccentric commutator	Check, regrind or turn.
	High mica	Undercut mica.
	Short in commutating pole turns	Check, repair or replace coils.
	Short in armature coils on commutator bars	Check and repair armature.
	Coils open circuited	Check and repair armature.
	Poor solder connection to commutator bars	Check and resolder.
	During high speeds bar high or loose at commutator	Check fasteners and tighten; grind face of commutator.
	Dirty commutator	Check and resurface.
	Vibrations	Check and correct.
Excessive brush wear	Soft brushes	Check and replace.
	Rough commutator	Grind commutator face.
	Not in neutral setting	Check for true neutral.
	Commutation poor	Check and correct.
	Bar loose, high, or low	Tighten commutator fasteners and resurface.
	Excessive brush tension	Reduce pressure.
	Film on commutator	Resurface brushes and commutator.

TABLE 5–9 (*continued*)

Problem	Probable Causes	Recommended Corrections
	Oil or grease contamination	Check, correct, and resurface brushes and commutator.
Noisy motor	Brush making singing or whistling noise	Check brush angle and commutator coating.
	Chattering brush	Resurface commutator and brush.
	Poor motor mounting or foundation	Check and correct.
	Belt bunching	Adjust tension.
	Bearings make noise	Check alignment, loading, and lubrication.
AC Motor Problems		
Stalling motor	Motor overloaded	Check motor rating and reduce load.
	Motor voltage too low	Increase voltage to motor rating.
	Improper use	Change motor size.
	Circuit open	Check fuse, circuit breaker, starter, and overload relay. Reset motor.
	Wrong control resistance of rotor	Check and replace resistors.
Motor does not start	Open circuit	Check and correct.
	Overload	Reduce load.
	One phase open	Check and close.
	Defective rotor	Replace.
Running motor dies down	Power supply off	Check line connections, fuses, breakers, and controls.

TABLE 5–9 (*continued*)

Problem	Probable Causes	Recommended Corrections
Motor fails to come up to speed	Wrong application	Check and correct.
	Too low voltage	Employ higher voltage.
	Starting load too great	Reduce load.
	Synchronous motor pull-in torque too low	Correct rotor starting resistance.
	Broken rotor bars	Check, repair or replace.
	In wound rotor the control operation of secondary resistance incorrect	Check and correct.
Excessive acceleration time	Excessive load	Reduce load.
	Poor or defective circuitry	Check for high resistance and correct.
	Squirrel cage rotor operation defective	Check and replace.
	Too low voltage	Increase voltage to proper level.
Incorrect rotation	Sequence of phases incorrectly installed	Reverse motor connections.
Overheating of motor while under load	Overloading	Check and reduce load.
	Poor ventilation	Check and correct restrictive condition.
	One of the motor's phases open	Check and connect all leads.
	Coil grounded	Check and correct.

TABLE 5–9 (*continued*)

Problem	Probable Causes	Recommended Corrections
	Terminal voltage not in balance	Check leads, connections, and transformers and make corrections.
	Stator coil shorted	Check and repair.
	High or low voltage	Check and correct.
Motor vibration	Motor not in alignment	Realign.
	Poor foundations or mounting	Check, tighten all fasteners, and strengthen base.
	Driven components out of balance	Check and correct.
	Bearing defective or not in line	Check, replace or align properly.
	Excessive end play	Adjust.
Noisy motor	Fan rubbing or striking air shield or insulation	Clear interference.
	Fan not firmly mounted	Tighten all mounting bolts.
	Loose bearings	Check, correct or renew.
	Unbalanced rotor	Check and rebalance.

MOTOR STARTERS

The proper maintenance of AC and DC motor starters is described in Table 5-10.

TABLE 5–10
Maintenance of Motor Starters

Maintenance Operation	Description/Comment
Replace contacts.	When contacts become worn, thinned, eroded, or badly pitted, replace them in pairs. Be sure to maintain correct pressures.
Keep contacts clean.	Keep contacts, especially copper contacts, clean by proper wiping, If the contacts become eroded, sand or file the surfaces.
Maintain tight connections.	Check fastening devices to be sure they hold contacts in place and are tight.
Lubricate according to manufacturer's specifications.	Do not oil contactor or relay bearings unless specified by the manufacturer. (Improper lubrication will cause accumulation of dirt and dust resulting in sluggish mechanical actions.)
Maintain correct voltage supply.	Operate coils at rated voltage. Measurement of operating voltage should be made at the coil terminals and not at the source of supply.
Maintain arc-rupturing components.	Keep arc-rupturing parts in good operating condition and in their correct operating position.
Replace frayed and worn shunts.	Replace frayed shunts at once to prevent overheating.
Maintain clean dashpots.	Be sure that oil dashpots have correct oil in them.
Take temperature readings.	Overheated parts indicate trouble; therefore, measure temperature if in doubt.
Check for unwanted grounding.	Eliminate undesired grounding on all circuits. If this is not done, there is potential for hazardous conditions, operating troubles, and erratic and dangerous operating circuits.

RHEOSTATS

Rheostats are used for starting and controlling the speed of DC and AC motors. Rheostat contacts should be kept in good operating condition by periodic inspection, cleaning, and dressing. Contacts should be smooth, and after each dressing, filed and polished with a fine abrasive paper or cloth. Contacting surfaces must then be thoroughly cleaned. The contacts may then be *lightly greased* with petroleum jelly or other similar lubricant.

Rheostats are often designed with reversible contacts that can be turned over and used on the reverse side when worn. This technique should only be used when there is excessive burning, making dressing ineffective. Pressure between moving and stationary contacts should be maintained to reduce heating, pitting, and corrosion.

SWITCH GEAR

Electric switch gears are used for power distribution and are commonly found in the form of circuit breakers and relays. Meters, instruments, switchboards, switches, and bus bars are examples of switch gear equipment. The frequency of inspection depends on the type of device, the operating and environmental conditions, and the method of installation.

Prior to any inspection or adjustment, make sure that the power has been turned off to all lines leading to the switch gear and that the device is completely disconnected. It is also a good practice to temporarily ground all breaker studs and current-carrying parts during inspection. Table 5-11 lists the recommended maintenance activities for various switch gear.

TABLE 5–11
Maintaining Switch Gear

Switch Gear Device	Maintenance Activity
Air circuit breakers	**Every 12 months:** Check and clean contacts. Make sure connections are intact, in good condition, and not overheating. Check that all fasteners are secure and tight. Check all mechanisms for binding or rubbing. See that all tripping devices and the trip latch move freely. Check calibration settings. Load current checks to make sure that they do not exceed ratings. Make sure that the operating voltage is sufficient. Check operation of control switch. Inspect mechanical condition of breaker enclosure.
Knife switches, high voltage fuses, and disconnect switches	**Every 12 months:** Check contacts with feeler gage; remove any oxide buildup, and check contact pressure. Check enclosed switches to make sure that the operating handle is functioning properly, that the fuse load is properly rated, and that ground connections are secure. Check outdoor high voltage fuses and clean contacts. Check high-voltage disconnecting switches and clean them.

TABLE 5–11 (*continued*)

Switch Gear Device	Maintenance Activity
Power circuit breakers	**Every 6 months:**
	Check oil for any moisture, carbonization, or dirt, and filter test.
	Every 12 months:
	Clean bushing surface, and check the oil level.
	Clean insulating and internal parts within the tank.
	Check operation of breaker mechanisms.
	Check contacts for alignment, pressure, and contact.
	Lubricate operating mechanisms.
Switch gear equipment	**Every 12 months:**
	Clean switchboards and enclosures.
	Examine and repair meters and instruments.
	Inspect the operation of control and instrument transfer switches.
	Replace any burned out indicating lamps.
	Check contacts of test blocks.
	Inspect bus bars and connection bars for condition, connections, and loads.
	Check primary and secondary connections for instrument transformers and potential transformer fuses.
	Inspect position-changing mechanisms for operation and lubricate.

TABLE 5–11 (*continued*)

Switch Gear Device	Maintenance Activity
	Check safety shutters and interlocks.
	Conduct dielectric tests if there has been any abnormal occurrence (e.g., fire, flood, storm damage).
Switch gear relays	**Every 6 months:**
	Check all high speed relays.
	Every 12 months:
	Check covers for broken or missing parts.
	Check and clean contacts.
	Check the operation of all moving parts.
	Inspect connections to make sure they are secure and intact.
	Inspect insulation for any deterioration.
	Check time settings.
	Conduct calibration and tripping tests.

TESTING EQUIPMENT

Once purchased and installed, electric testing and metering devices usually require minimum maintenance. However, they should be periodically checked, inspected, and calibrated for proper function, output, and readings. Occasional replacement of expendable parts is considered routine maintenance.

Preventive Maintenance of Testing Equipment

1. Keep the piece of equipment clean and dry.
2. Maintain the apparatus to ensure reliability, efficiency, and maximum service life.
3. Maintain cleanliness in the work area to minimize the chance of flashovers from high voltage circuits and power supplies.
4. Conduct scheduled inspections to determine maintenance needs. Examples of what to look for include overheating, cable or lead displacement, loose connections, and oxidation or corrosion of conductors.
5. Restrict adjustments and alignments to restoring proper output, performance, and readings.
6. Remember that lubrication is usually not needed, except where bearings, motors, or shafts are involved.
7. Check switches for positive contact and proper operation. Replace worn, erratic, or defective switching devices and interlocks.
8. Replace defective equipment or return it to the supplier or manufacturer for repair.
9. Inspect meters that have broken protective glass and replace.

TRANSFORMERS

The maintenance of transformers requires regular inspections and checks, along with a running repair and replacement program. The design of these programs varies according to the type of transformer. Some procedures require checks several times a day; others can be scheduled once every five to seven years. Table 5–12 gives a recommended maintenance schedule for various types of transformers and their components.

TABLE 5–12
Maintenance of Transformers

Type of Transformer or Component	Maintenance Operation	Frequency
All transformers	Check liquid level gage, ambient temperature, and liquid and winding temperature indicators to make sure they are within specified limits.	3 times a day
	Check load current on transformer banks.	
	Check relays.	
	Check overvoltage protective equipment.	4 times a year
	Check ground connections and resistance to ground.	
Askarel transformers	Check relief diaphragm for cracks and test askarel.	4 times a year
	Check gas absorbers for damp or caked compounds.	
	Check the condition of askarel.	2 times a year
	Check tanks by subjecting them to internal pressures of 5 psi (pounds per square inch) for 12 hours.	
	Check under the cover for condensation.	Every 5 years
	Overall inspection.	Every 7–10 years
Dry-type transformers	Check fan operation.	3 times a day

TABLE 5–12 (*continued*)

Type of Transformer or Component	Maintenance Operation	Frequency
	Check core and coils for dust accumulation and corrosion.	4 times a year
	Check temperature alarm.	
	Check fan lubrication and contact surfaces.	Once a year
Forced oil cooled transformers	Check pump glands and oil level.	3 times a day
	Check ingoing and outcoming oil temperature.	Once a week
	Check cooler screens for any foreign material.	
	Check oil strainer for clogging.	
Gas-seal equipment	Check gas pressure and adjust gas regulator.	3 times a day
	Check gas cylinder for gas depletion.	Once a month
	Check pressure relief valve and minimum pressure alarm circuit.	4 times a year
	Check gas regulator.	
	Check oxygen content. (There should not be more than 5 percent nitrogen content.)	2 times a year
Indoor air blast transformers	Check air temperatures of incoming and outgoing air.	Once an hour
	Blow out transformer with dry compressed air.	Once a week

TABLE 5–12 (*continued*)

Type of Transformer or Component	Maintenance Operation	Frequency
Load-ratio-control device	Check relief diaphragm.	3 times a day
	Check adjustments and operation of voltage regulating relay, auxiliary relays, fuses, and other similar components.	4 times a year
	Filter oil and clean insulating surfaces.	Once a year
	Check contactors.	
	Inspect arcing ratio adjustors.	
	Lightly lubricate stuffing boxes.	
	Check mechanisms.	
	Inspect ventilators.	
	Check insulating liquids.	
Oil conservator, gas seal, gas-oil seal, and sealed transformers	Check gas pressure relief diaphragm for cracking or breaks.	3 times a day
	Check pressure vacuum gage.	
	Check for moisture in the insulating liquid.	2 times a year
	Check oil level on idle or spare transformers to make sure air does not empty conservator.	
	Perform pressure tests for leaks above the oil level.	Once a year

TABLE 5–12 (*continued*)

Type of Transformer or Component	Maintenance Operation	Frequency
	Check above the core for condition.	Every 5 years
	Overall inspection.	Every 7 years
Oil immersed air pressure cooled transformers	Check oil temperature.	3 times a day
	Check fan operation and inspect automatic controls of motor.	Once a month
Open-type transformers	Check breathing to make sure ventilators are free of obstructions.	4 times a year
	Check the condition of the oil.	
	Check under the cover for moisture.	2 times a year
	Check above the core for oil sludge deposits and moisture.	Every 2 years
	Inspect overall transformer and internal parts.	Every 5 years
Water cooled transformers	Record ingoing and outgoing water temperatures.	Once a week
	Check cooling coils.	
	Check water pressure and flow for clogging.	2 times a year

6 Mechanical Equipment

The maintenance of mechanical equipment is the most common area of responsibility for the machine and maintenance mechanic. This chapter discusses maintenance and diagnostic procedures for the component parts of most industrial and commercial machinery.

BEARINGS

Bearings are designed to support rotating or oscillating parts and protect them from damage. Bearings should be selected on the basis of their ability to provide low friction, low wear, conformability, and embedability. When subjected to abnormal conditions, the bearing material should yield rather than distort or damage the supporting member.

Bearings are made from a variety of materials, with tin-base and lead-base babbit considered the best all-around material. Other metals, such as bronzes, copper leads, cadmium-base, aluminum-base, and silver are also used. Recently, other materials, such as polymers and ceramics, have also been employed as bearing materials.

The major consideration in maintaining any type of bearing is to follow those procedures recommended by the manufacturer. Detailed assistance and replacement consultation is often available from the nearest office of the bearing manufacturer.

CATEGORIES OF BEARINGS

Bearings can be grouped into two major categories: fluid-film bearings and rolling contact bearings.

Fluid-Film Bearings. Fluid-film bearings make up a broad category of bearings. These bearings are divided into three subgroups:

1. *Journal Bearings,* which are cylindrical in shape and are designed to carry rotating shafts;
2. *Thrust Bearings,* which are designed to prevent any lengthwise movement of rotating shafts;
3. *Guide Bearings,* which are specifically used to guide machine elements in a lengthwise motion. Usually, these elements will not have any rotational movement.

Rolling Contact Bearings. Rolling contact bearings are used to support and rotate shafts. Perhaps the most recognizable rolling contact bearings are ball and roller bearings. Figure 6-1 shows examples of common types of rolling contact bearings.

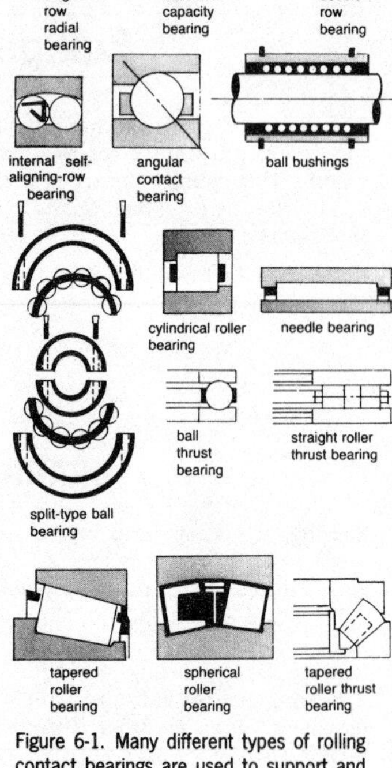

Figure 6-1. Many different types of rolling contact bearings are used to support and rotate shafts.

CLASSIFICATION OF BEARINGS

Bearings should be selected based on the type of duty they will perform and other factors. To address this need all bearings are available in several standard classifications: 100, 200, 300, and 400 series. (This series is comparable to an older designation technique, in which light, medium, and heavy-duty designators were used—comparable to the 200, 300, and 400 series respectively.)

LUBRICATION OF BEARINGS

Cleaning and proper lubrication is fundamental to proper ball and bearing operation. A well-lubricated bearing will operate indefinitely until repeated stresses and strains start subsurface cracking that eventually leads to failure. Lubrication is applied to bearings to support the contact between the bearing surface and the moving part. It also protects the high finish of the bearing surface from corrosion, high speed temperatures, and foreign material. Table 6-1 presents lubrication considerations for bearings.

TABLE 6–1
Lubrication of Bearings

Lubrication Consideration	Description
Oil lubrication	The type of oil selected should be based on operating speeds, bearing load, and ambient temperatures. Lower limits for each factor require lower oil viscosity ratings; while higher limits dictate higher oil viscosity ratings. Check manufacturer's recommendations.
Grease lubrication	Bearing greases are usually a mixture of lubricating oil and a soap base. When operating temperatures are below 32°F, greases with a lithium soap as a gelling agent should be used; above 32°F, agents such as soda or soda lime should be used. Check manufacturer's lubrication charts for frequency of greasing.

TABLE 6–1 (*continued*)

Lubrication Consideration	Description
Moisture protection	Sodium-base greases form a noncorrosive film when exposed to moisture. When moisture is present, grease should be replenished. Excessive water flow dictates the use of a nonemulsifying and water-repellent grease.
Idle machinery protection	Equipment not in full service should be set in motion periodically to allow the spread of lubricant over all bearing surfaces. This should be done every 30 to 90 days, as dictated by atmospheric conditions.

TROUBLESHOOTING BEARING PROBLEMS

Many of the problems associated with bearings in various kinds of machinery can be prevented by proper maintenance or can be quickly solved—if the problem is recognized. Table 6-2 lists some common problems, their probable causes, and recommended solutions.

TABLE 6–2
Troubleshooting Bearing Problems

Problem	Probable Causes	Recommended Corrections
High running temperature	Overlubrication	Check oil level: it should be at the centerline of the lowest bearing when not rotating. One fourth of the housing should be filled with lubricant.

TABLE 6–2 (*continued*)

Problem	Probable Causes	Recommended Corrections
	Excessively tight shaft or housing fittings	Check shaft seat and housing bore for diameter, taper, and roundness against manufacturer's specifications.
	Misalignment of shaft or housing	Check shaft and housing dimensions and centerlines against manufacturer's specifications.
Noise, vibrations, or rough running	Assembly damage	Check and repair or replace.
	Foreign material between bearings	Inspect and clean bearings with solvent. Replace lubricant with high quality clean oil or grease.
	High-speed starts with insufficient or congealed lubricant	Check lubricant level and replenish if low. If congealing is the cause, replace lubricant with recommended grade and quality.
	Overlubrication	See above.
	Excessively tight shaft or housing fittings	See above.
	Misalignment of shaft or housing	See above.
Fatigue failure	Poor lubrication	Check lubrication levels periodically.

TABLE 6–2 (*continued*)

Problem	Probable Causes	Recommended Corrections
Discoloration of bearings	Overheating and metal smearing	Check, clean with solvent, and properly lubricate.
	Extended operation beyond rating	Check and correct.
Excessive wear	Dirt or abrading foreign material	Check and clean.
	Incorrect bearing clearance	Check measurements against manufacturer's specifications and correct.
Corrosion	Moisture	Check type of lubrication being used and be sure that it is a nonemulsifying and water-repellent grease.
	Chemical reaction caused by incorrect lubricant	Lubricant should be noncorrosive emulsion mix.

BELT DRIVES

Much industrial and commerical equipment still makes extensive use of belts and accessories such as fasteners, pulleys, and shafts for the transmission and conversion of power. The most important aspect of these power transmission components is their design and installation.

Belts should be made of appropriate materials and construction. They should be of correct size and duty for handling required horsepower, speed, and pulley size; should be tough enough to withstand applied loads; and should be able to maintain lacing and hole fasteners effectively.

Fasteners, regardless of the material that they are made out of, have to be able to withstand belt speeds and other working conditions.

The pulley and shaft also have to be of the correct material and size to handle the required horsepower, speed, load, and bearing support.

COMMON BELTING MATERIALS

Common belting materials used for power transmission are leather, rubber, stitched cotton or canvas, and balata. Environmental and operating conditions determine the type of belting to be used. Belting materials recommended for specific operating conditions are as follows:

Acid contact: camel's hair– solid woven belting.

Acidic mists and vapors: mineral retanned leather, rubber, and camel's hair– solid woven belting.

Abrasive dusts: all common materials except rubber-covered belting.

Steam and moisture: all common materials except nonwaterproof oak leather.

Temperatures to 115°F: all common belting materials.

Temperatures to 130°F: all common belting materials except nonwaterproof oak leather and friction surface rubber.

Temperatures to 140°F: only mineral retanned leather, solid woven cotton, or camel's hair– solid woven.

Temperatures over 140°F: camel's hair– solid woven.

Mineral oils and grease: all common materials, except rubber and rubber-covered belting.

Normal conditions: all common belting materials.

BELT MAINTENANCE AND TROUBLESHOOTING

Regularly scheduled inspection of belting and accessories will prevent many problems. Table 6-3 summarizes the maintenance considerations for such power transmission systems. Table 6-4 provides hints for troubleshooting belt problems.

TABLE 6–3
Maintenance of Belt Drives

Drive Component	Maintenance Procedure
Belt	Inspect for dryness, brittleness, dirt, and oil saturation, surface glazing, lap and joint condition, tension, and shaft-pulley alignment.
Fastening method	If endless, check for any looseness or curling at the lap joint and for unevenness; correct running direction of the lap and wind.
	Metallic fasteners should be checked for corrosion and wear, proper size and type, and any interference with the belt.
Pulley	Inspect for effects of environment such as temperature, moisture exposure, and other severe conditions.
	Check for balance, trueness of diameter, atmospheric effects, and alignment. Remove any areas that have sharp edges or crowning.
Shaft	Inspect for straightness, alignment, rigidity, bearing support, and fastening.

TABLE 6–4
Troubleshooting Belt Problems

Problem	Probable Causes	Recommended Corrections
Slipping and squealing belt	Loose belt	Increase tension of belt.
	Overload for belt capacity	Replace belt with heavier duty belt or reduce load.
	Belt surface too dry or glazed	Check and apply proper dressing or replace belt.
	Crown of the pulley too high	Check and reduce crown taper.

TABLE 6–4 (*continued*)

Problem	Probable Causes	Recommended Corrections
Excessive stretching of belting	Incorrect belt size and/ or material	Check and replace with belt that has appropriate capacity or is made of proper material.
Belt not running true	Belt stretching on one side or not squared when joined; misaligned pulleys, or running onto flanged or step-cone pulley	Repair or replace belt, and correct installation problem.
Belt coming off pulley when running	Pulleys or shafting misaligned	Check and correct alignment; replace or repair belt.
	Belt not running true	See above.
Belt running along one side of pulley	Belt too loose	Increase belt tension.
	Excessive load	Reduce load or use higher capacity belt.
	Belt not running true	Repair or replace belt.
	Pulley or shafting misaligned	Check and correct.
Belt flapping or whipping during operation	Motor and parts not secured, bent shaft, pulley lopsided, variating load at the power source, or incorrect belt tension	Check and correct. When possible, attempt to change motor speed or use fly wheel to smooth out the load.
Crack in the belt	Incorrect belt tension	Check and correct.
	Diameter of pulley undersized	Check and replace.
	Overheating of belt caused by slippage	Increase belt tension.

CHAIN DRIVES

Assuming that the precision chain drive was properly selected and installed, its sprockets correctly installed, and sufficient lubrication provided over time, it is then possible to list the maintenance procedures that should be followed.

Maintenance Procedures for Chain Drives

1. Check the chain drive periodically for alignment. Misalignment is judged when the sides of the sprocket teeth or the inside faces of the chain link plates show sign of wear. When such wear is evident, realignment should be executed immediately.

2. Check the chain itself for any excessive slack. Chains that run close to the larger sprocket's teeth tip should be replaced. Excessive slackness can be checked when the drive is off by pulling the chain away from the larger sprocket to make a visual inspection. (See Figure 6-2.) The chain should be in alignment with the sprockets. When there is too much clearance, no amount of tension adjustment will keep the two in alignment, and chain replacement is required.

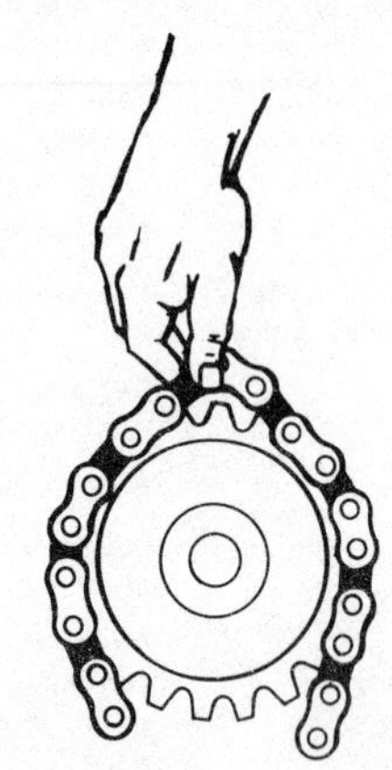

Figure 6-2. Pull the chain away from the sprockets to check for excessive slackness.

3. If the sprockets are worn, do not install a new chain, for it will cause excessive and premature wear. (It should be noted that the life of a sprocket can be extended by reversing it on the shaft.)

4. Whenever a new drive is installed, inspect it frequently for any interference with the chain. If rubbing or striking does occur, replacement will be necessary.

5. Check sprocket teeth for the presence of foreign material that will cause the chain to ride high. Such a condition will cause excessive stresses and wear.

6. Lubrication checks should be made for grade and quality of lubricant. Four common lubrication systems and their maintenance requirements are:

   Manual lubrication, which should be checked regarding lubrication schedule and the proper application of oil.

   Drip Lubrication, which requires the inspection of oiler cups and feed rate. Also make sure that the feed pipes are not clogged.

   Bath or *disk lubrication,* which should be checked for oil level and to make sure that there is no sludge. These systems should be drained, flushed, and refilled at least once a year.

   Force-fed lubrication systems, which should be inspected for reservoir level, pump drive and delivery pressure, and sludge. Again, these systems should be drained, flushed, and refilled at least once a year.

7. Notice that whenever improper or insufficient lubrication of the roller chains exists, the joints have a brownish color and the pins of the connecting link of the chain are also discolored. The pins may also be grooved, roughened, or galled.

8. Clean the chain periodically to ensure maximum operating efficiency. First, remove the chain from the sprockets and wash it in kerosene. If it is excessively dirty and gummed, soak it for several hours until the material is loosened. Next, drain the cleaning fluid off and soak the chain in oil; this will restore the internal lubrication of the metal. Then hang the chain to drain off any excess oil. Finally, inspect the chain for any wear or corrosion. During chain cleaning, it is good practice to also inspect the sprocket and to clean it with kerosene.

9. If the chain drive is idle, store it to minimize deterioration. This can be accomplished by removing the chain and sprockets, coating them with heavy oil or light grease, and storing them in grease-resistant paper. If the sprocket is to be left on the shaft, it is good practice to cover it to prevent damage. Before placing the drive back into service, thoroughly clean the chain and sprockets and relubricate the chain.

CHAIN HOISTS

Chain hoists are either hand or electrically operated. Because of their simple design, ease of use, and relative low cost, chain hoists have become standard equipment in many repair shops.

TYPES OF CHAIN HOISTS

There are five basic types of chain hoists, each suited to a particular use.

Differential Hoist. The differential hoist is the least expensive of all chain hoists and has a simple design that provides for easy lifting by a differential in the diameters of two pocketed grooves in the upper sheave.

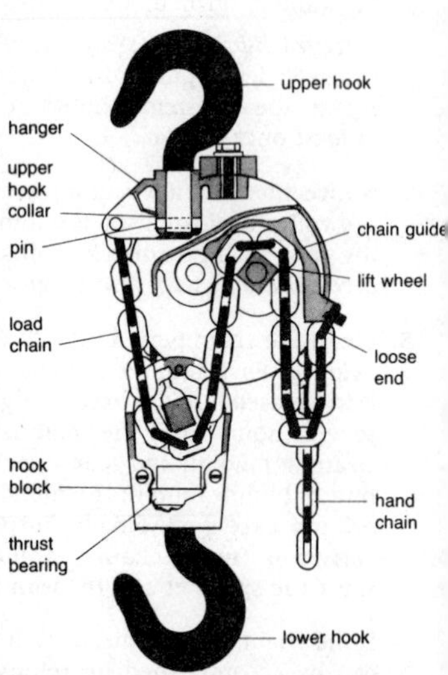

Figure 6-3. Parts of a chain hoist should be inspected and serviced regularly.

Screw-Gear Hoist. A screw-gear hoist incorporates the worm and gear principle and is approximately twice as efficient as a differential hoist. Although slower than the differential design, it is of great advantage when smoother lifting is required.

Spur-Gear Hoist. Although it is more expensive than the preceding two hoists, a spur-gear hoist is much more efficient. Friction is reduced so much that a braking device is necessary to hold the load in place. There are three basic spur-gear hoist designs: twin hook, extended hand wheel, and low headroom trolley hoists.

Puller. A puller, which is a simple form of chain hoist, is operated by a ratchet lever that is capable of lifting, pulling, dragging, or stretching loads in any direction.

Electric Hoist. An electric hoist is a spur-gear hoist that has a push-button or pendant-rope control. The switching mechanism has three motor controls: up, down, and stop. Electric hoists come in various sizes, ranging from small capacity models equipped with 115 volt motors to heavier duty models with dual 220/440 volt power motors.

MAINTENANCE OF CHAIN HOISTS

The design and construction of modern chain hoists has kept required maintenance to a minimum. Figure 6-3 illustrates hoist parts that should be inspected and serviced regularly. Table 6-5 presents a recommended maintenance schedule for chain hoists.

TABLE 6–5
Maintenance of Chain Hoists

Hoist Part	Procedure	Frequency
All Hoists		
Overall hoist	Operate while not under load and check for free and smooth operation and running of chains.	Once a month
Hand chain and handwheel	Clean chain and wheel pockets. Inspect for wear, stretched links, or other damage.	Once a month
Housing, frame, and covers	Inspect for loose fasteners and cracks or other signs of damage. Make sure that gaskets are tight and not leaking lubricant.	Once a month
Load brake	Follow general friction brake maintenance procedures. Some brakes require no lubrication so check manufacturer's recommendations.	Once a month

TABLE 6–5 (*continued*)

Hoist Part	Procedure	Frequency
	Disassemble and clean all parts. Check ratchet teeth and pawl tip and bore for wear, and replace if necessary. Inspect friction surfaces for wear. Remove all foreign material, and reassemble.	Once a year
Load chain	Clean and inspect for wear, damage, and stretched links. Lubricate chain with wipe of penetrating oil or graphite.	Once a month
Lower hook	Inspect for stretching, wear, and damage. Inspect to assure free swiveling. Check condition of bearings, nuts, collars, and pins.	Once a month
	Lubricate as required.	
	Double reeved hoists require disassembly of hook block for inspection of any wear.	Once a month
Upper hook	Check for stretching, wear, and damage. Inspect to ensure free swiveling, and good condition of bearings, nuts, collars, and pins. Lubricate as necessary.	Once a month
Electric Hoists		
Electric brake	Operate under load and check for holding power, drift, and drag. Adjust friction members as required. Lubricate cam, cam followers, and linkage sparingly.	Once a month

TABLE 6–5 (*continued*)

Hoist Part	Procedure	Frequency
Electrical connections and insulation	Tighten screw terminals and securely crimp all wires. Inspect insulation.	Once a year
Limit switches	Inspect and clean operating linkages, contact points, and terminals.	Once a year
Magnetic contactor	Inspect and clean terminals, operating coils, and line contacts.	Once a year
Motor	Most modern motors require no more maintenance than general cleaning. For older models, follow manufacturer's recommendations.	Once a year

FRICTION CLUTCHES

Friction clutches are widely used machine elements that enable one rotating element to transmit torque to another element by means of friction. Clutches come in two basic designs: one in which the driving and driven shafts are coaxial and run at the same rate of speed; the other, a sleeve-type clutch, in which one element of the clutch functions as a sleeve in which a separate wheel is attached.

If friction clutches are properly selected and installed, maintenance needs will be minimal. The following practices are recommended:

Maintenance of Friction Clutches

1. Institute a suitable lubrication program to avoid bearing problems.

2. Check friction clutches that have just been installed or are infrequently used daily for signs of overheating. In some cases, it is recommended that lubricants be added frequently, even though this may result in excess lubrication.
3. Inspect parts subjected to heavier loads, such as bushed sleeve bearings, more frequently than other components, such as ball and roller bearings.
4. Keep clutches clean and well lubricated to avoid operating slippage and failure.

Problems with friction clutches usually involve the running parts and manifest themselves by excessive heating or wear. Table 6-6 gives hints for recognizing and correcting common friction clutch problems.

TABLE 6–6
Troubleshooting Friction Clutches

Problems	Probable Causes	Recommended Corrections
Overheating of bearings	Inadequate lubrication	Check and replenish lubricant to recommended levels.
	Dirt and other foreign material	Disassemble, clean, and replenish lubricant.
Excessive slippage	Insufficient torque capacity	Change friction clutch to proper rating.
	Change in load	Apply additional power. (Replacement may be necessary.)
Loss of capacity	Friction surfaces contaminated by oil or grease	Disassemble and clean with solvent.
	Excessive wear of parts	Repair or replace parts and lubricate according to manufacturer's specifications.

TABLE 6–6 (*continued*)

Problems	Probable Causes	Recommended Corrections
Failure to engage fully	Loose adjustments or excessive wear caused by foreign obstruction or corrosion of sliding parts	Thoroughly clean and lubricate. If excessive wear, replacement may be necessary.
Insufficient slip ring travel	Lost motion	Inspect air lines and quick release valves for any clogging.

GEAR DRIVES AND SPEED REDUCERS

Machine elements widely used to change speed, torque, and rotational direction include gear drives and speed reducers. These elements use sets of gears mounted on shafts and enclosed in casings that make use of gaskets, oil seals, and air breathers. Common types of industrial gear drives are spur, helical, herringbone, plain bevel, spiral bevel, hypoid, worm, rack and pinion, and internal. (See Figure 6-4.)

Basic gear drives are employed to transmit, change, or modify power between a central mechanical power supply and a driven machine. Drives are usually installed as a total unit and are manufactured according to standards set by the American Gear Manufacturers Association (AGMA). Perhaps the most common drive product is the motorized gear drive, which is used throughout many industrial facilities.

Central to a good maintenance program is the use of quality lubricating oils within the enclosed unit.

Lubricating Gear Drives

1. Select a neutral lubricant that will not corrode the gearing and ball or roller bearings.

2. There should be no grit or abrading material within the gearing unit. If any is detected, remove it immediately. Then clean the gear drive unit with a solvent and replenish with a clean and high quality lubricant.

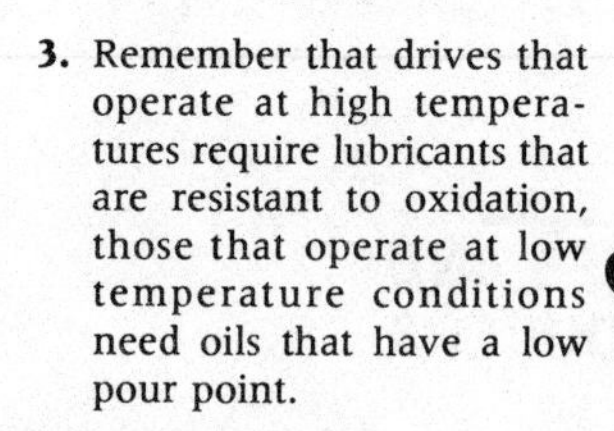

3. Remember that drives that operate at high temperatures require lubricants that are resistant to oxidation, those that operate at low temperature conditions need oils that have a low pour point.

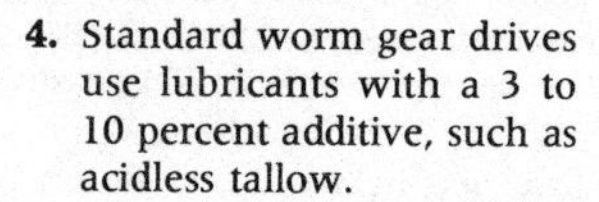

4. Standard worm gear drives use lubricants with a 3 to 10 percent additive, such as acidless tallow.

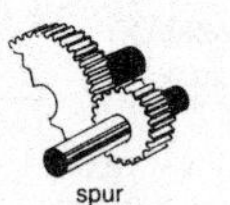

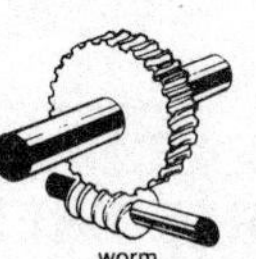

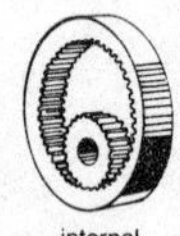

Figure 6-4. Many different types of industrial gear drives are used to transmit, change, and modify power between a central power supply and a driven machine.

5. Gearing subjected to extreme shocks or impact loading should use an extreme-pressure (EP) lubricant.

6. Check pressure fittings for leakage frequently.

7. Lubricate bearings at scheduled intervals.

Table 6-7 provides hints for solving problems involving gear drives and speed reducers. Table 6-8 deals specifically with adjustable pulleys.

TABLE 6–7

Troubleshooting Gear Drives and Speed Reducers

Problems	Probable Causes	Recommended Corrections
Wear	Normal wear caused by abrasion during normal operation	Inspect periodically and lubricate properly.
	Abrasive wear caused by foreign particles	Stop the unit, drain oil, clean entire unit, and replenish with clean lubricant.
	Large particle scratching	Same as for abrasive wear. Protect against recontamination by using filters and breathers.
	Overloading unit	Reduce unit load.
Plastic yielding	Heavy, uneven loads and sliding	Check, reduce backlash of gear teeth, and use extreme pressure lubricant.
	Excessive high temperatures	Check and use an extreme pressure lubricant. Consult gear manufacturer.
Surface fatigue	Surface irregularities	Grind and polish tooth bearing surface for initial pitting. Destructive pitting requires the additional practice of using extreme pressure lubricant.
Corrosive wear	Acid or alkaline contamination	Drain and flush gear case, and remove contaminant.
Burning	Overloading, overspeeding, lack of backlash, or poor lubrication	Check lubricant and be sure that gears are not underrated for load and speed.

TABLE 6–7 (*continued*)

Problems	Probable Causes	Recommended Corrections
Interference	Heavy contact between gear teeth	Check for improper design, manufacture, deflection, or assembly, and consult gear manufacturer.
Tooth breakage	Overloading	Reduce load.
	Fatigue	Reduce cycles or revolutions of applied loads.

TABLE 6–8
Troubleshooting Variable Speed Drives (Adjustable Pulley)

Trouble	Probable Causes	Recommended Corrections
Stationary Control		
Rapid belt wear	Misalignment of driver and driven pulleys	Realign pulleys.
	Continuous flexing over small diameter	Select driven pulley of proper diameter.
	Excessive heat, cold, humidity, acidic atmosphere, abrasives, and other environmental conditions	Check and correct.
	Overloading, excessive shock loads, or excessive belt speeds	Check and correct.
	One or a few belts taking the load in a multiple belt pulley system	Rematch belt lengths more closely.
	Belt tension is too high	Adjust the center distance.

TABLE 6–8 (*continued*)

Trouble	Probable Causes	Recommended Corrections
Belt slippage	Greasy pulley faces	Clean pulley faces.
	Overloading	Adjust loads.
Controlled in Motion		
Excessive belt wear	See above	See above.
Belt slippage	See above	See above.
Belt not running level	Improper or insufficient lubrication of spring-loaded disks	Stop motor, disassemble, clean, and lubricate. Maintain regular lubrication schedule.
Belt Transmission		
Excessive belt wear	Disk assembly out of alignment	Check and realign so that constant speed and variable disk assemblies are parallel at average speed.
	See above	See above.
Belt slippage	Greasy pulley faces	Check and clean.
	Constant speed shaft set too slow	Check and increase input speed.
	Insufficient belt tension	Check and increase belt tension.
Belt makes a creaking sound	Belt tension excessive	Check and reduce belt tension.
Bearing failure	Excessive belt tension	Check and reduce belt tension.
	Overhung load exceeds capacity	Check and correct.
	Poor lubrication practices	Check and follow manufacturer's lubrication specifications.

TABLE 6–8 (*continued*)

Trouble	Probable Causes	Recommended Corrections
	Abrading or corrosive atmospheric conditions	Enclose belt transmission drive.
	Incorrectly assembled unit or bent shaft	Check and correct.
Speeds cannot be adjusted	Improper lubrication of disks, causing them to stick	Stop use of machine, disassemble, clean disk hub and shaft with solvent, and lubricate according to manufacturer's recommendations.
All Metal Traction Systems		
Not capable of shifting speeds May shift slightly, but will slide back to original setting	Sliding surfaces locked in position due to operating extensively at one speed	Replace damaged parts; may need to return to manufacturer.
Thumping sound	Load change or overloading	Replace damaged parts; may need to return to manufacturer.
Excessive overheating	Insufficient lubrication	Check and replenish to recommended levels.
Excessive slippage	Lubricant weight too heavy	Flush, clean, and replenish with proper viscosity lubricant.
Flat Belt		
Belt slippage	Overloading	Reduce load.
	Stretched belt	Replace belt or increase distance between pulleys or use an idler roll.
	Greasy pulley faces	Check and clean.

7

Service Equipment

The successful operation of any industrial or commercial facility depends on the type of services or support systems that are available. Since these systems often do not directly affect the productivity and operation of a facility, they are sometimes neglected for more "important" jobs. Machine and maintenance mechanics who understand the importance of service equipment appreciate the need for a proper maintenance program, which, in turn, minimizes problems that may affect the efficiency of the facility. This chapter describes maintenance and troubleshooting procedures recommended for typical service systems and equipment.

AIR CONDITIONING SYSTEMS

The term "air conditioning" is usually synonymous with cooling systems, but in reality, air conditioning pertains to both heating and cooling systems. Although the two systems are completely different in their objective, they often use the same ventilation system (e.g., motors, fans, and ducts). In fact, a significant number of facilities make use of the same ductwork and mechanisms for both cooling and heating.

Three factors must be kept in mind in maintaining all types of air conditioning equipment: cleanliness, tightness, and the effective operation of safety devices. Cleanliness is critical if heat transfer, fluid flow, and proper lubrication is to be maintained. Tightness is important because it reduces the loss of capacity and refrigerant and the ingress of contaminants. Safety devices ensure that the system will not destroy itself or cause harm to individuals working in its environment.

COOLING SYSTEMS

Most cooling systems or plants fall into one of three general categories:

1. *Reciprocating plants* that use a reciprocating compressor, which may be from 1/6 horsepower (e.g., residential refrigerators) to over 400 horsepower units
2. *Centrifugal plants* that employ centrifugal compressors with capacities between 100 and 2,500 tons per unit
3. *Absorption plants* in which the cooling is produced when one fluid is absorbed by another. Engineers favor absorption machines because the power requirements of these systems are limited to circulating the fluids within the unit. In addition, these units are advantageous where there are large amounts of water and waste steam or heat twelve months a year.

MAINTAINING COOLING SYSTEMS

Air conditioning assemblies follow no fixed arrangement, and in fact, the names of component parts often vary from one manufacturer to the next. Thus, there is no one maintenance sequence that fits all situations. Table 7-1 presents a maintenance schedule suitable for most commercial air conditioning cooling units.

TABLE 7–1

Maintenance Schedule for Air Conditioning Cooling Equipment

Equipment	Maintenance Procedure	Frequency
Air compressor	Align.	During installation
	Lubricate.	Annually
	Adjust belt tension and alignment.	Monthly
	Blow out drip pockets and eliminators.	As needed
	Clean out air take.	Annually

TABLE 7–1 (*continued*)

Equipment	Maintenance Procedure	Frequency
	Clean filters.	Seasonally
	Check and replenish oil chambers.	Annually
	Inspect screens.	Weekly
Sprays	Inspect strainers and tanks.	Weekly
	Check and adjust controls.	Annually
	Perform general inspection.	Annually
	Check oil level and replenish as needed.	Monthly
Air washer	Inspect strainers and tank.	Weekly
	Check and adjust controls.	Seasonally
	Provide freeze protection.	Seasonally
	Inspect for leakages and repair.	Annually
	Perform general inspection.	Annually
	Check operating pressures.	Weekly
	Inspect for spray erosion.	Monthly
	Check pH level of water.	Weekly
Apparatus room	Clean.	Weekly
	Provide freeze protection.	Seasonally
	Perform general inspection.	Weekly

TABLE 7–1 (*continued*)

Equipment	Maintenance Procedure	Frequency
Circuit breakers or starter	Lubricate.	Annually
	Inspect strainers and tanks.	Annually
	Check controls.	Annually
	Perform general inspection.	Annually
	Check oil level.	Annually
Coils	Clean off dust.	Seasonally
	Check drainage.	Seasonally
	Provide freeze protection.	Seasonally
	Perform general inspection.	Annually
	Check for leakage and repair.	Annually
	Pump down.	Seasonally
Condensers	Check drainage.	Seasonally
	Provide freeze protection.	Seasonally
	Perform general inspection.	Annually
	Inspect for leakages and repair.	Annually
	Check operating pressure.	Daily
	Check pH level of water.	Weekly
Conditioned space	Check and adjust settings and calibrations.	Weekly
	Perform general inspection.	Weekly

TABLE 7–1 (*continued*)

Equipment	Maintenance Procedure	Frequency
Controls	Blow out drip pockets and eliminators.	Weekly
	Clean.	Annually
	Check and adjust settings and calibrations.	Weekly
	Inspect protective devices.	Monthly
	Perform general inspection.	Seasonally
Coolers	Clean off dust.	Seasonally
	Check and adjust controls.	Monthly
	Check drainage.	Seasonally
	Provide freeze protection.	During installation
	Perform general inspection.	Seasonally
	Check for leakage and repair.	Annually
	Check operating pressures.	Daily
	Pump down.	Seasonally
	Check level of refrigerant.	Weekly
	Check pH level of water.	Weekly
Cooling towers	Lubricate.	Monthly
	Check for overheating and correct.	Seasonally
	Inspect oil chambers.	Seasonally
	Inspect screens.	During installation

TABLE 7–1 (*continued*)

Equipment	Maintenance Procedure	Frequency
	Inspect sprays.	Seasonally
	Inspect strainers and tanks.	Weekly
	Check for effective operation.	Monthly
	Provide freeze protection.	Seasonally
	Perform general inspection.	Seasonally
	Check for leakages and repair.	Annually
	Check oil level and replenish.	Seasonally
	Check operating pressures.	Monthly
	Inspect for spray erosion.	Seasonally
	Check stand-by.	Seasonally
	Check pH level of water.	Weekly
Dampers	Check alignment.	Seasonally
	Lubricate.	Annually
	Check controls and adjust.	Seasonally
	Inspect for effective operation.	Monthly
	Perform general inspection.	Annually
Dehumidifiers	Check sprays.	Monthly
	Inspect strainers and tanks.	Weekly
	Check controls and adjust.	Seasonally
	Provide freeze protection.	Seasonally

TABLE 7–1 (*continued*)

Equipment	Maintenance Procedure	Frequency
	Perform general inspection.	Annually
	Check for leakages and repair.	Annually
	Check oil pressures.	Weekly
	Check pH level of water.	Weekly
Distribution system	Perform general inspection.	Annually
	Clean off dust.	Annually
Drives	Check alignment.	Monthly
	Check belt tension and alignment.	Monthly
	Perform general inspection.	Annually
Evaporative condenser	Lubricate.	Seasonally
	Check for overheating.	Seasonally
	Inspect oil chambers.	Weekly
	Inspect sprays.	Seasonally
	Inspect strainers and tanks.	Weekly
	Provide freeze protection.	Seasonally
	Perform general inspection.	Seasonally
	Check for leakages and repair.	Annually
	Checking operating pressures.	Weekly
	Check for spray erosion.	Monthly
Fans	Check alignment.	During installation
	Lubricate.	Monthly

TABLE 7–1 (*continued*)

Equipment	Maintenance Procedure	Frequency
	Check for overheating.	Seasonally
	Check and adjust belt tension and alignment.	Monthly
	Check oil chambers.	Annually
	Perform general inspection.	Annually
	Check oil level and replenish.	Monthly
	Rotate fans between units.	Annually
Filters	Blow out drip pockets and eliminators.	Annually
	Clean.	Annually
	Check oil chambers.	Seasonally
	Check strainers and tanks.	Seasonally
	Check controls and adjust.	Monthly
	Check for effective operation.	Monthly
	Perform general inspection.	Annually
	Check oil level and replenish.	Monthly
Freeze protection	Check controls.	Seasonally
	Check drainage.	Seasonally
	Perform general inspection.	Seasonally
Heating coils	Check drainage.	Seasonally
	Provide freeze protection.	Seasonally

TABLE 7–1 (*continued*)

Equipment	Maintenance Procedure	Frequency
	Perform general inspection.	Annually
	Check for leakages and repair.	Annually
Humidifiers	Check sprays.	Monthly
	Inspect strainers and tanks.	Weekly
	Check controls and adjust.	Seasonally
	Provide freeze protection.	Seasonally
	Perform general inspection.	Annually
	Check for leakages and repair.	Annually
	Check operating pressures.	Weekly
	Check for spray erosion.	Monthly
	Check pH level of water.	Weekly
	Check calcium deposits.	Monthly
Insulation	Perform general inspection.	Annually
Lint screens	Blow out drip pockets and eliminators.	Seasonally
	Clean.	Daily
	Perform general inspection.	Annually
Motors	Lubricate.	Monthly
	Check for overheating.	Seasonally
	Check belt tension and alignment.	Monthly

TABLE 7–1 (*continued*)

Equipment	Maintenance Procedure	Frequency
	Clean off dust.	Seasonally
	Check oil chambers.	Annually
	Check strainers and tanks.	Annually
	Perform general inspection.	Annually
	Check oil level and replenish.	Monthly
	Rotate usage of motors.	Annually
Operating conditions	Perform general inspection.	Daily
Pumps	Check alignment.	Annually
	Lubricate.	Monthly
	Check for overheating.	Seasonally
	Check filters and clean or replace.	Annually
	Check strainers, screens, and tanks.	Seasonally
	Provide freeze protection.	Seasonally
	Perform general inspection.	
	Check oil level and replenish.	Monthly
	Check stand-by.	Seasonally
	Check seals.	Annually
Refrigerant piping	Check for leakage.	Monthly
	Pump down.	Seasonally
Refrigerant compressor	Check alignment.	Seasonally
	Lubricate.	Monthly

TABLE 7–1 (*continued*)

Equipment	Maintenance Procedure	Frequency
	Check for overheating.	Monthly
	Check belt tension and alignment.	Monthly
	Check oil chambers.	Seasonally
	Check controls and adjust.	Monthly
	Perform general inspection.	Annually
	Check oil level and pressure	Daily
	Check operating pressure.	Daily
	Pump down.	Seasonally
	Check stand-by.	Seasonally
	Inspect seals.	Weekly
Room air conditioner	Inspect filters.	Weekly
	Inspect strainers and tanks.	Seasonally
	Inspect controls and adjust.	Seasonally
	Perform general inspection.	Seasonally
Self-contained units	Check and adjust belt tension and alignment.	Monthly
	Check filters and clean.	Seasonally
	Check controls and adjust.	Seasonally
	Perform general inspection.	Seasonally
	Check operating pressure.	Monthly

TABLE 7–1 (*continued*)

Equipment	Maintenance Procedure	Frequency
	Check for spray erosion.	Seasonally
Steam piping	Blow out drip pockets and eliminators.	Seasonally
	Check strainers and tanks.	Annually
	Perform general inspection.	Annually
	Check for leakages and repair.	Monthly
Water conditioning	Check strainers and tanks.	Weekly
	Inspect for effective operation.	Weekly
	Perform general inspection.	Weekly
Water piping	Blow out drip pockets and eliminators.	Seasonally
	Check sprays.	Weekly
	Check drainage.	Seasonally
	Provide freeze protection.	Seasonally
	Perform general inspection.	Annually
	Check for spray erosion.	Seasonally
	Check pH level of water.	Weekly

TROUBLESHOOTING COOLING SYSTEMS

Troubleshooting any type of cooling system involves eight activities:

1. Measuring the temperature of each cooled area or space

2. Determining the operational status of each control thermostat
3. Measuring the temperature from discharge of the air handler
4. Determining the refrigerant suction pressure and temperature
5. Noting the temperature and pressure of the liquid as it leaves the condenser
6. Checking condenser conditions
7. Determining the length of run for compressor
8. Checking to see if there is any unusual sound.

A more detailed troubleshooting checklist for commercial cooling systems is given in Table 7-2.

TABLE 7–2
Troubleshooting Cooling (Mechanical Commercial) Systems

Problem	Probable Causes	Recommended Corrections
Mechanical Commercial Systems		
Area to be cooled is too warm	Dirty condenser coils and air filter	Check and remove all dust and foreign substances.
	Low refrigerant level	Take reading and replenish to correct level.
	Evaporator coils frosted	Inspect and defrost.
	Strainer and expansion valve dirty	Check, clean or replace as needed.
	Loose or improperly positioned thermostatic bulb (on coil)	Check and correct.
	High head pressure	Locate cause of excessive pressure and correct.

TABLE 7–2 (*continued*)

Problem	Probable Causes	Recommended Corrections
	Incorrect condensing water temperatures	Compare temperatures to manufacturer's manual. Check condition of water, and clean out.
	Noncondensable gas or air in the system	Purge system and recharge.
Noisy compressor	Refrigerant entering compressor	Locate refrigerant leak and correct.
	Worn pistons, out-of-round cylinders, loose wrist pins, broken piston rings, and/or defective valves	Take compressor apart and repair or replace defective parts.
	Loose foundation or other supports	Tighten all hold-down bolts, keyways, and/or shaft screws.
Compressor does not start	No power	Check and correct.
	Blown fuse, burned out or defective switch, and/or burned out capacitor	Check electrical system and replace defective components.
	Bearings overheated and insufficiently lubricated, causing crankshaft to freeze	Test shaft seal for binding and inspect unit internally. Repair or replace bearings, housing, and crankshaft.
Compressor motor has burned out several times	Undesired voltage fluctuations	Check for low voltage and correct.

TABLE 7–2 (*continued*)

Problem	Probable Causes	Recommended Corrections
	Unit struck by lightning	Install lightning arrestors.
	If hermetically sealed, contaminants present	Check for acids and sludge or other contaminants.
Hissing sound by expansion valve, and inadequate cooling	Partial blockage of expansion valve	Clean valve and strainer or replace as required.
	Defective orifice or undersized expansion valve	Check and replace.
Conditioned areas are too hot or too cold	Faulty zone balancing	Test air volume delivered with volumeter, and adjust zone dampers for proper balancing.
	Defective thermostats	Check and replace.
Excessively noisy system	Undersized registers and air ducts	Lower volume or temperature.
	Misaligned shaft seal	Replace or repair.
	Low crankcase oil level	Check oil level and add additional oil to proper level
	Omitted or undersized expansion valve	Check and correct.
Unpleasant odors coming from the system	Short charge of refrigeration	Check and correct.
	Dirty air filters	Clean or replace filters. Clean ducts.

TABLE 7–2 (*continued*)

Problem	Probable Causes	Recommended Corrections
	Decomposing organic material within the duct	Locate and remove.
	Source of odor originating from outside the building	Relocate fresh air intake unit and/or treat with activated carbon.
Commercial Absorption Systems		
System not performing up to rating	Insufficient heat supply to generator	Increase steam pressure.
	Dirty and sluggish heat exchangers between condenser and generator	Clean exchangers.
	Condenser supplying insufficient amount of cooled water	Clean pipelines and cooling tower.
	Water pumps not circulating water as rated	Check and repair pumps.
	Refrigerant absorbent out of balance	Add absorbent or refrigerant to bring back into balance.

HEATING SYSTEMS

Heating may be accomplished directly or indirectly. In direct systems—direct fired heaters—the air to be heated passes directly over the energy source. In indirect systems the combustion energy heats some other material—for example, steam or water—which then conveys the heat to a given area. In both cases the heating system provides a means for:

1. Moving warmed air from the heating surface or area to the space to be heated
2. Guiding the air to be heated over a heated surface
3. Providing a heated surface
4. Transfering energy from one source so that it can heat the heating surface
5. Supplying a source of energy to the heating surface.

TYPES OF HEATING SYSTEMS

Within most industrial and commercial operations, there are several types of heating units available. A brief description of each follows:

1. *Direct Fired Units* are typically found in central heating facilities, packaged unit heaters, and suspended unit heaters. They are designed in a variety of sizes and shapes.
2. *Built-up Units* are either brick set (like a boiler) or steel cased units used for central heating and incorporated into the design and layout of the building. They come in capacities from 400,000 to 800,000 Btu per hour.
3. *Packaged Heating Units* are factory assembled units with oil, gas, or combination gas-oil burners. Mobile, flexible, and easy to install, these units provide for an alternative heating source. They come in sizes from 200,000 to 2,500,000 Btu per hour.
4. *Suspended Unit Heaters* burn either natural gas or light fuel oil and have a capacity range of 25,000 to 200,000 Btu per hour. Though they are lightweight, flexible for small areas, and easily installed, suspended unit heaters do not keep uniform temperatures at the work level, have excessive hot air blowing in the area of the unit, consume significant amounts of fuel, and are often difficult to maintain.
5. *Steam Heating Systems* are either high pressure, low pressure, vacuum, or vapor systems. (The latter is not commonly used in industrial and commercial settings.) The proper design and maintenance of the pipe system that transports the steam is critical.

6. *Hot Water Heating Systems* pipe heated water to heat dispersing units. These systems are classified by the type of circulation used (that is, gravity or forced) or by the piping design (that is, one pipe, two pipe, direct return system, or two pipe reverse return system).

MAINTAINING HEATING SYSTEMS

Maintenance requirements for heating systems are usually incorporated into those activities suggested for cooling systems. A major concern is regular inspection and care of heat transporting devices such as piping, valves, ducts, and motor-fan units. For a discussion of boiler maintenance, see Chapter 4; for a discussion of proper maintenance procedures for cooling systems, power plant equipment, and ventilating exhaust systems, see the appropriate sections in this chapter.

TROUBLESHOOTING HEATING SYSTEMS

There are a large number of heating system designs, each with its own maintenance requirements and special problems. Table 7-3 provides hints for troubleshooting common heating systems.

TABLE 7–3
Troubleshooting Heating Systems

Problem	Probable Causes	Recommended Corrections
One or more rooms fail to heat in a gravity system.	Supply registers are at the end of a long duct or leader.	Install larger duct, increase duct insulation, and/or raise temperature.
	Warm air ducts are incorrectly pitched.	Check to be sure all supply lines are pitched in an upward direction.

TABLE 7–3 (*continued*)

Problem	Probable Causes	Recommended Corrections
	Return registers are too high above the floor.	Install active return register and/or provide for additional heating source.
	The outside air intake damper has been opened too widely.	Inspect and either close or throttle the intake damper.
	Entire system is unbalanced—some rooms are too hot.	Check all dampers and reset to bring system into balance.
Registers' dirt and dust are settling throughout the facility.	Ducts are dirty.	Clean duct system with a vacuum.
	Airways around the furnace radiator are excessively dirty.	Check, wipe, and vacuum.
Floor furnace heats only immediate areas.	There is poor warm air flow.	Provide an additional auxiliary heating source. Use additional fans to circulate air.
Smudging is found around diffusers and registers in forced air system.	Ducts, grills, and registers have excessive dirt.	Vacuum ducts and all outlets.
	Air filter is dirty.	Clean or replace air filter.
	Room is very dusty and circulates dirt within the room when the system is operating.	Clean room on a regular basis.

TABLE 7–3 (*continued*)

Problem	Probable Causes	Recommended Corrections
System is noisy.	Air flow through the system is at too high a velocity.	Check and reduce system air velocity.
	Blower and/or other machinery are vibrating, which is transmitted through the ducts.	Set on vibration absorbers or install insulating absorbers around ducts.
There are undesirable odors in heated areas.	Excessive dirt, dust, and other deposits are in ducts.	Vacuum ducts.
	Decaying organic material is present in the duct system.	Locate and remove material.
	Odors originate from outside the building.	Trace source; if odor is obnoxious, install activated carbon purifier.
Furnace runs continually without sufficient heating.	Undersized heater, and/or system is unbalanced.	Undersized heaters demand use of an auxiliary peak heater. Correct imbalance.
	Heated area is drafty.	Install weather stripping and storm sash.
	Ceilings are high.	Use auxiliary heater for improved balance.
There is rattle noise during fan operation.	Bearing is malfunctioning and/or fan is not secure on shaft.	Replace bearing and tighten fan.
Area cools when doors are closed.	Area has inadequate heating grills.	Provide openings at bottom of doors (e.g., louvers).

TABLE 7–3 (*continued*)

Problem	Probable Causes	Recommended Corrections
	Thermostat location is not related to heated area.	Relocate thermostat or provide for openings at bottom of doors.
Air is too dry.	There is insufficient water vapor in building.	Increase relative humidity by installing properly rated humidifier.
	Dry outside air is entering building.	Check and correct with proper insulation.
	Lack of humidifier.	Install humidifier.
	Defective humidifier.	Check and correct.
Air is so humid that condensation forms on windows.	The structure was built "tight" with effective vapor barrier.	Install ventilation fan or dehumidifier.
Smoke appears from warm air registers when heater is fired by gas or oil.	Heat exchanger is leaking or corroded.	Repair by welding or replace.
	Dust or dirt near the heater is burning.	Clean filter and heater components.
	Material settled in ducts is combustible.	Use vacuum to draw material out. Clean ducts. If fire is present, call the fire department immediately.

CENTRIFUGAL PUMPS

Centrifugal pumps are used to move fluid. They are the most common type of pump in use today. Made from a number of common metals and metal alloys as well as from porcelain, glass, and polymers, these pumps have one or more rotating impellers in a fixed casing that guide fluid to and from the impeller or from one impeller to the next. Impellers may be either single or double suction.

INSTALLING CENTRIFUGAL PUMPS

Successful maintenance of centrifugal pumps begins with proper selection and installation. Centrifugal pumps should be installed in a location that is accessible for normal inspection and as near the fluid supply as possible to allow for short and direct suction. Pumps are accurately aligned by the manufacturer; the same amount of care should be taken at the site. When connecting pipes to the pump, make sure that they align naturally. Avoid forcing the pipes into place with flange bolts. The discharge pipe should never be smaller than the pump discharge. Also make sure that the V-belt drive is perfectly aligned to avoid excessive belt wear. (See Figure 7-1.)

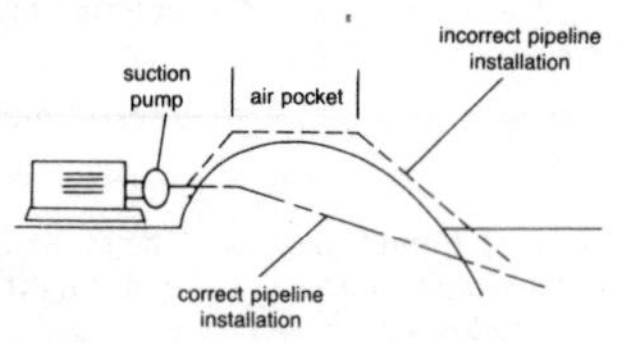

Figure 7-1. Piping should be connected to a centrifrugal pump so that it aligns naturally.

MAINTAINING CENTRIFUGAL PUMPS

Once installed, make sure that the piping has been properly aligned. The pump must rotate in the direction indicated by an arrow on the casing. When starting the pump, first prime it so that the suction pipe is completely filled.

Over the life of the pump, it is important to institute a proper lubrication program for all bearings. The mechanic should also make sure that the bearing housing is kept very clean and that all foreign solids and liquids within the housing are removed.

Wearing-ring clearances should be checked periodically to make sure that they are within operating limits. Excessive wear can be caused by fluids containing gritty and/or corrosive contaminants. Shaft sleeves should be replaced when leakage cannot be controlled by normal tightening pressures.

Pumps should be packed with long fiber material; square braided packing that is impregnated with oil and graphite. When the transported liquid is not water, special packing may be required. In recent years, mechanical seals have gained in popularity and are rapidly replacing packing. One of the major advantages of mechanical seals is that they can provide absolute control of leakages.

TROUBLESHOOTING CENTRIFUGAL PUMPS

Centrifugal pumps that have been properly maintained seldom break down. When they do, the problems usually occur in the form of water not being delivered or insufficient pumping pressures. Presented in Table 7-4 are the causes and corrections for common pump problems.

TABLE 7–4

Troubleshooting Centrifugal Pump Problems

Problem	Probable Causes	Recommended Corrections
No water being pumped	Pump not primed	Completely prime pump.
	Speed of the pump too low	Check to see if correct voltage is being supplied.
		For steam turbine check governor.
		Check motor size to discharge size for accuracy.
		Check operating conditions and correct.
	Discharge head elevated too high.	Check operating conditions and correct.
	Suction lift elevated too high.	Check gages and correct.
	Plugged impeller and/or line	Check pipe line; suction strainer and impeller.
	Impeller rotating in wrong direction	Check that rotation is in the direction away from the vane curvature.

TABLE 7–4 (*continued*)

Problem	Probable Causes	Recommended Corrections
Insufficient amount of water being delivered	Air leaks in suction line and/or stuffing box	Inspect and make necessary repairs or adjustments.
	Too slow speed	Check starting directions in manufacturer's manuals.
	Discharge head higher than expected	Check operating conditions and make necessary corrections.
	Suction lift too high	Check with gages and adjust.
	Partially plugged impeller or suction line	Inspect and clean out obstruction.
	Insufficient suction for hot liquid	Refer to pump manuals.
	Worn wearing rings	Check and replace rings.
	Damaged impeller	Check and either repair or replace as necessary.
	Undersize foot valve	Inspect and replace.
	Defective casing packing	Inspect and replace if worn.
	Foot valve and/or suction opening not submerged	Check to see that suction line is at least three feet below the surface of the liquid.
Insufficient pressure	Speed too low	Check for proper line voltage, or with steam turbine governor, for sufficient pressure.
	Air in the liquid	Check for leaks.
	Worn wearing rings	Inspect and replace.

TABLE 7–4 (*continued*)

Problem	Probable Causes	Recommended Corrections
	Damaged impeller	Check and repair or replace as required.
	Defective casing packing	Inspect and replace all worn packing.
Suction lost after the pump works for a short time	Leak in the suction line	Check for leaks and correct.
	Plugged water seal	Inspect line and location of the seal cage in the stuffing box and remove foreign material.
	Suction lift too high	Check to see that the recommended height of the suction lift does not exceed manufacturer's recommendations.
	Air or gas in the liquid	Check line for leaks.
Excessive power consumption by pump	Speed set too high for pump	Check driver, pulley or sheave diameters.
	Head rating too low for work	Check and replace with higher capacity pump.
	Mechanical damage	Check shaft and other components for defects such as being bent or out of round.
	Binding rotating elements	Inspect stuffing boxes, wearing ring fit, and packing for defects or excessive tightness.

DUST COLLECTING EQUIPMENT

There are two broad categories of air cleaning equipment: air filters and dust collectors. The first is primarily used for air supply systems, such as for heating, air conditioning, ventilation, and compressor intakes. Dust collectors are designed to handle greater concentrations of particles. Generally, dust collectors are designed to handle 100 to 20,000 times more particles than are air filters.

TYPES OF DUST COLLECTING DEVICES

Dust collecting designs vary according to manufacturer and application. Most, however, can be classified according to their operating mechanism (See Table 7-5).

TABLE 7–5
Types of Dust Collecting Equipment

Operating Class	Types of Collectors
Inertial or dry centrifugal	Simple cyclone Multiple centrifugal Dry dynamic Louver
Wet	Packed tower Wet centrifugal Wet dynamic Orifice design Nozzle design Venturi scrubber
Fabric arrest	Conventional Reverse jet
Electrostatic	

MAINTAINING DUST COLLECTING SYSTEMS

As with other service equipment, proper maintenance begins with good design, selection, and installation. The basic factors involved in maintaining dust collecting equipment are cleaning hoppers regularly and at predetermined scheduled intervals; instituting an inspection and preventive maintenance program; and repairing, replacing, and cleaning parts as needed.

Table 7-6 lists maintenance practices for various types of dust collecting systems.

TABLE 7–6
Maintenance of Dust Collecting Equipment

Equipment	Maintenance Procedures
Inertial or Dry Centrifugal	
Simple cyclone and multiple centrifugal	Empty storage hoppers at regular intervals to prevent reentrainment.
	Inspect and repair any leakages at the dust discharge point, particularly in units under negative pressure.
	Provide scheduled inspections for buildup and indications of excessive wear.
	To maximize efficiency, maintain uniform air volume.
Dry dynamic	Maintain venting of dust storage hopper through to the collector.
	Regularly inspect impeller for accumulation and wear along the blades.
	Maximize unit efficiency by regularly emptying the dust storage hopper.
	Follow maintenance practices for fans, bearings, and belts.
Wet Collectors	
Wet centrifugal and wet dynamic	During fan operation, make sure that there is an adequate supply of water.

TABLE 7–6 (*continued*)

Equipment	Maintenance Procedures
	Make sure that systems using recirculated water are equipped with settling tanks to ensure clear water.
	After the fan is shut down, allow water to flow through the system for 30 minutes to provide adequate flushing.
	Schedule regular inspections for wear and leakage, and thoroughly clean as required.
	For systems using spray nozzles, inspect for clogging.
	Inspect drains for accumulations and plugging.
Fabric Arresters	
Conventional and reverse jet	Periodically inspect for any dust leakage through holes or openings in the fabric.
	Check for excessive wear or holes in baffle plates and spark screens.
	Regularly empty dust storage hoppers and check for excessive accumulations.
	Inspect dust discharge valves for proper operation.
	Provide regular inspections of fabric rapping devices to ensure proper cleaning action.
	Inspect jet ring and secondary air supply of reverse jet units.
Electrostatic	
Precipitators	For mechanical components designed with rapping mechanism, follow same procedures as given for fabric arresters.

TABLE 7–6 (*continued*)

Equipment	Maintenance Procedures
	Follow proper grounding procedures for all electrical maintenance inspections.
	Check ionizing and grounded plate section for accumulations of dust and grit.
	Inspect dust outlets and storage hoppers.
	Test electrical system for proper ionization.
Sludge setting tanks	Remove sludge from tanks before they become filled.
	Keep the clean water chamber free of silt.
	In units with conveyors inspect links and paddles.
	Inspect hopper bottom for excessive accumulation.
	Check chain guides and hopper wear plates for excessive wear.
	Check for proper tension in conveyor and adjust as necessary.
	Check pumps for abrasion and wear.
Fans	Check fan mountings to avoid undesired vibrations.
	Inspect impeller disk for abrasive wear and corrosion.
	Check for proper fan rotation.
	Make sure that V-belts have proper tension.
	Schedule routine lubrication of bearings.

FLUID POWER SYSTEMS

Fluid power systems are used in a wide range of machines found in manufacturing plants, job shops, and maintenance repair shops. These systems use flowing fluids to provide power and movement. Fluid power systems are broken down into two basic areas: hydraulics and pneumatics.

HYDRAULIC SYSTEMS

Hydraulics, a segment of fluid mechanics, uses oil and other liquids. Within machinery, the liquid is used under pressure for power transmission and in control devices such as servomechanisms. The basic hydraulic system consists of four parts:

1. The *pump*, which moves the oil or liquid.
2. The *cylinder*, which makes use of the moving oil to accomplish work.
3. *Check valves*, which hold oil within the cylinders between strokes and prevent it from returning to the reservoir while under pressure (pressure stroke). These valves are designed to open while the liquid is flowing and to close while it is at rest.
4. A *reservoir*, which is a storage tank for the oil or liquid. All reservoirs have an air vent to allow the oil to be fed into the pump by gravity or atmospheric pressure.

MAINTAINING HYDRAULIC SYSTEMS

Within most industrial settings, the maintenance of hydraulic systems requires well planned activities. Poor maintenance can result in inadequate oil in the reservoir, clogged and dirty filters, loose intake lines, use of an improper oil grade, and incorrect operating pressures.

Good hydraulic maintenance programs can be broken down into three simple considerations:

1. Maintain a clean and sufficient quantity of hydraulic fluid of the proper type and viscosity. The first fluid change should be made 50 to 100 hours after start-up. Thereafter, fluid change should take place every 5,000 hours on small systems and after a maximum of 10,000 hours on large systems.
2. Change filters regularly and clean strainers. During start-up, filters should be checked every two to three hours. Afterwards, daily checks should be made, and after seven days of operation, cleaning may be required.
3. Be sure that all connections are tight to keep air out of the system. However, connections should not be tightened to the point of distortion. Piping or tube systems should be inspected regularly for leakage. Main and pilot pressures should be checked once a week and adjustments noted.

PNEUMATIC SYSTEMS

In pneumatic systems, air flow is used for power transmission and controlling devices. The basic pneumatic system consists of three parts:

1. The *compressor*, which is used to convert mechanical energy transmitted by a motor into compressed air.
2. The *heat exchanger*, also known as an *aftercooler*, which is used to cool the compressor-discharged air. In most cases, the cooling medium is water, though there are some air-to-air exchangers for large compressors.
3. The *storage tank*, which holds the compressed air and serves as an energy accumulator. Another function of the storage tank is to dampen severe pulsations.

MAINTAINING PNEUMATIC SYSTEMS

The maintenance of pneumatic systems also includes maintaining air tools. Generally the efficient operation of pneumatic systems depends on three critical concerns: air lines, shutoff couplers, and lubrication.

1. *Air lines* supply the tool or machine with compressed air, and they should have the correct inside diameter. The basic rule of thumb is that the diameter should be one size larger than the inlet thread.
2. Self-sealing or automatic shutoff *couplers* are usually employed for connecting the air line to the tool or machine. Couplers function as a valve for the air supply and are easily snapped together for a tight connection. Care should be taken to use the correct capacity coupler for the tool at hand, so that proper operating pressure is maintained.
3. *Lubrication* is very important. The vast majority of tool and machine failures can be traced to inadequate lubrication. Machinery and larger tools have a built-in reservoir for oil, but it is important to check that the machinery is actually receiving the oil from the line.

TROUBLESHOOTING FLUID POWER SYSTEMS

Although troubleshooting and correcting problems in fluid power systems often requires specialized training, the machine and maintenance mechanic should have some knowledge of common fluid power system problems and how they can be remedied. (See Table 7-7.)

TABLE 7–7
Troubleshooting Fluid Power System Problems

Problem	Probable Causes	Recommended Corrections
Excessive noise in pump	Gaseous fluid in a liquid stream (Cavitation)	Replace dirty filters; wash strainers in solvent; clean clogging in inlet line; clean reservoir breather vent; change fluid; repair or replace supercharge pump; use properly rated pump drive motor speed; or increase fluid temperature.

TABLE 7–7 (*continued*)

Problem	Probable Causes	Recommended Corrections
	Air in the fluid	Check and tighten inlet connections; bleed air from system, replace pump shaft seal if worn; or fill reservoir to correct level.
	Misalignment of coupling	Check and realign unit.
	Worn or damaged pump	Inspect and overhaul or replace pump.
Noisy motor	Coupling not properly aligned	Inspect seals, bearings, and coupling. Realign unit.
	Worn or damaged coupling and/or motor	Check and either repair or replace.
Excessive noise from valve	Valve setting too low or close to another valve	Check with pressure gage and adjust.
High fluid temperature	System pressure too high	Check with pressure gage and adjust.
	Fluid dirty, of incorrect viscosity, or at inadequate level	Change filters and fluid if improper. Fill reservoir.
	Fluid cooling system malfunctioning	Inspect and either repair or replace.
	Worn or damaged pump, valve, motor, cylinder, or other device	Check and repair or replace.
Excessive pump heat	Cavitation, air in fluid, or worn or damaged pump	See above.

TABLE 7–7 (*continued*)

Problem	Probable Causes	Recommended Corrections
	High fluid temperature	See above.
	Excessive load	Check if load is proper for system's circuit design. Correct any mechanical binding.
	Unloading and/or relief valve set too high	Check with pressure gage and correct.
Motor overheating	High fluid temperature	See above.
	Excessive load	See above.
	Worn or damaged motor	Check, overhaul or replace.
	Unloading or relief valves set too high	See above.
Overheated relief valve	High fluid temperature	See above.
	Incorrect valve setting	Check with pressure gage and correct.
	Worn or damaged relief valve	Check, and repair or replace.
No fluid flow in the circuit	No fluid reaching the pump	Replace dirty filters; clean clogged inlet lines; clean reservoir breather vent; repair or replace supercharge pump; and/or fill reservoir to correct level.
	Nonoperative pump drive motor or improperly assembled pump	Check, repair or replace.
	Damaged pump	Repair or replace.

TABLE 7–7 (*continued*)

Problem	Probable Causes	Recommended Corrections
	All of the flow passing over the relief valve	Test and adjust.
	Directional control incorrectly set	Check controls and electrical circuit on solenoid controls. Repair or replace pressure pump.
	Coupling between pump and drive broken	Replace.
	Pump drive motor turning in wrong direction	Check and correct.
Low flow rate	Flow control, relief, or unloading valve set too low	Test and adjust.
	Inoperative yoke-actuating device; worn pump, valve, motor, or cylinder	Check and repair or replace.
	Flow being by-passed through a partially opened valve	Check controls and adjust, repair, or replace valve.
Excessively high flow rate	Incorrect rpm of pump drive, or incorrect size pump	Replace with correctly rated unit.
	Flow control incorrectly set	Check and adjust.
	Yoke-actuating device inoperative	See above.
No pressure	No flow	See above.

TABLE 7–7 (*continued*)

Problem	Probable Causes	Recommended Corrections
Low pressure	Damaged pressure-reducing valve, pump, motor, or cylinder, or worn relief valve	Check, repair or replace.
Erratic circuit pressure	Fluid contamination	Replace dirty filters and fluid.
	Air in fluid	Check for leaks and tighten all connections.
	Worn relief valve, pump, motor, or cylinder	Repair or replace.
	Defective accumulator or loss of charge	Tighten all connections and check for leakages.
Excessive pressure	Yoke-actuating device inoperative; or worn or damaged pressure reducing, relief, or unloading valves	Repair or replace.
	Incorrectly adjusted pressure reducing, relief, or unloading valves	Check and correct.
No movement in system	No flow or pressure	See above.
	Worn or damaged cylinder or motor; inoperative servo, or inoperative or incorrectly adjusted limit or sequence device.	Repair or replace.
	Mechanical bind	Find location of bind and repair.

TABLE 7–7 (*continued*)

Problem	Probable Causes	Recommended Corrections
	Incorrectly adjusted or inoperative servo-amplifier	Adjust, repair, or replace.
Slow movement of fluid in the system	Fluid viscosity too high for design of circuit	Check fluid temperature and/or replace with correct viscosity fluid.
	Low flow	See above.
	Insufficient control valve pressure	See above.
	Worn or damaged motor or cylinder	See above.
	Poorly lubricated machine ways and/ or linkages.	Inspect and lubricate.
	Inoperative, misadjusted, or malfunctioning servo amplifier	See above.
	Servo-valve sticking	Clean and adjust; check condition of fluid and filters.
Erratic system movement	Erratic pressure	See above.
	Air in fluid	See above.
	Worn or damaged cylinder or motor; malfunctioning or incorrectly adjusted servo-amplifier	Check, repair, or replace counterbalance valve, or adjust properly.
	Poorly lubricated machine ways and/ or linkages	Check and lubricate.
	Command signal erratic	Test and repair command console or interconnecting wires.

TABLE 7–7 (*continued*)

Problem	Probable Causes	Recommended Corrections
Excessively high system speed or movement	Excessive flow	See overleaf.
	Malfunction in the feedback transducer	Check and repair or replace.
	Servo-amplifier malfunctioning or incorrectly adjusted	See above.
	Overriding of the work load	Inspect, test, repair, or replace counterbalance valve.

POWER PLANT EQUIPMENT

Power plant equipment is an important element of most industrial and commercial facilities. It is designed to provide various forms of power required for day-to-day operations. In most cases, the equipment is designed not only to provide the energy needed to make processing possible but also to contribute to the air conditioning of the plant.

The maintenance of power plant systems is designed to anticipate and prevent shortcomings and major repairs, or, in other words, to minimize repair by maximizing upkeep. Hence, power plant maintenance falls into two broad categories: breakdown and planned maintenance.

A practice, unique to power plant maintenance, is that of *shutdown operations*. This is a preplanned process in which the entire facility is shut down, or closed, for a period of one to two weeks so that maintenance operations can be performed. Due to the obvious cost of such procedures, shutdown operations are usually scheduled at most once per year, and in many situations, only every two to five years.

FUEL-BURNING EQUIPMENT

The maintenance of fuel burning equipment is primarily concerned with corrosive, abrasive, and other mechanical damage that may occur. Boiler maintenance, often considered a part of power plant maintenance, is discussed in Chapter 4. Equipment commonly associated with fuel burning and maintenance considerations are listed here.

1. *Coal bunkers* are used to store coal. An important maintenance consideration is the possible presence of corrosive coals, such as those with a high sulfur content. To minimize damage, empty the bunker completely and clean it before it is refilled with a new supply of coal.
2. *Coal chutes* are used to move coal from one location to another. Maintenance concerns include not only corrosive attack but also abrasion caused by the movement of coal along the chute. Most steel chutes have a service life of about six to eight years.
3. *Pulverizing equipment* reduces coal to a more efficient size, by impact, attrition, and/or crushing. The common types found in power plants are ball, impact, ring-roll, and ball-race pulverizers. There is little maintenance on pulverizing equipment. Factors to observe during operation include maintaining correct charge size, mill temperature, and coal level.
4. *Mechanical stokers* are used primarily to produce steam for small and moderate-sized generators. Maintenance mechanics should ensure loading within approved limits, maintain the proper fuel draft, prevent damage to the grates caused by fires in the wind box, maintain proper banking methods, and lubricate as needed. All visible and accessible parts of the stoker should be inspected at least once a day.
5. *Fuel oil burning equipment* is commonly used in industrial settings to produce steam. Proper maintenance should include the preparation of fuel oil before it reaches the burner, the inspection of the oil storage tank, scheduled cleaning of sludge and the removal of moisture, checking heating coils for steam leakage, and proper cleaning and replacement of burner tips.

6. *Gas burning equipment* is the most efficient way to produce steam. Critical here is the prevention of unwanted gas leakages which can cause serious accidents. Checks of burner cock vents, gas regulating controls, and the main supply line should be made regularly.

STEAM-GENERATING EQUIPMENT

Steam is widely used in both industrial and commercial power plants. Although the responsibility for maintaining this equipment falls on mechanics who have received specialized training, it is important for all mechanics to understand proper maintenance procedures. Table 7-8 lists typical maintenance procedures for steam-generating equipment.

TABLE 7–8

Maintenance of Steam Producing Equipment

Equipment	Maintenance Procedure
Turbine generator	Listen for any rubbing or binding noises from the bearings or clicking noises at governor housing and oil pump within the casing.
	Check and maintain correct oil level in storage tank. Look for moisture and proper feed for return oil.
	Inspect packing for excessive leakage.
	Check throttle valve operation; oil all links; clean all exposed stem threads; and lubricate bearings.
	Check governor for movement of levers and linkages.
	Take generator temperature periodically. Check collector rings and exciter commutator for sparking, heat, and busing vibration.
Economizers or heat recovery units.	Frequently check steam operating pressures.

TABLE 7–8 (*continued*)

Equipment	Maintenance Procedure
	Check the pH level of feed water; check for any signs of oxidation.
Air preheaters	Check and clean all surface areas exposed to flue gases.
	Control corrosion of these surfaces.
Induced and forced draft equipment	Clean blades of fly ash and inspect for proper fastening.
	Check fan alignment and operations during hot and cold conditions.
	Check rotor balance at full speed.
	Check bearings, oil reservoirs, and water elements for water cooled bearings.
	Inspect fan housing, connecting ducts, and control dampers.
Surface condensers	Check and regularly clean tubes.
	Check for any deficiency in cooling water.
	Inspect for tube and air leakage.
	Check for faulty air removal.
Mechanic drive turbines	Inspect bearings, oil rings, and oil level periodically.
	Check for packing leakage.
	Inspect proper operation of governor, throttle valve, governor valve, and relief valve.
	Clean nozzle plate and stationary and rotating buckets.
De-aerators	Overhaul and clean entire unit at least once a year.
	Check storage tanks for corrosion and paint them.

TABLE 7–8 (*continued*)

Equipment	Maintenance Procedure
Strainers	Clean all baskets regularly.
	Check seal fitting for basket.
	Keep record of all maintenance procedures followed and any changes in fluid being strained.
Steam traps	Test on a regularly scheduled basis.
	Open traps at least once a year to inspect parts and clean.

TROUBLESHOOTING STEAM GENERATING EQUIPMENT

Small steam turbines are used in many industrial facilities to generate power. To ensure efficient, problem-free use, it is important that the turbines be correctly installed and maintained. Maintenance schedules should include inspections of critical parts, bearings, oil rings, oil level, packing leakage, governor, and the throttle valve every eight hours of operation.

Table 7-9 provides a diagnostic list of possible problems involving mechanic drive steam turbines, their causes, and recommended corrections. This table is a general guide; for more detailed explanations, consult the manufacturer's manual.

TABLE 7–9
Troubleshooting Mechanic Drive Steam Turbines

Problem	Probable Causes	Recommended Corrections
Operating at excessive speeds	Sticking of the governor valve stem or trip weight	Clean valve stem and lubricate; smooth weight and clean spring.
	Governor incorrectly adjusted	Check and adjust trigger.

TABLE 7–9 (*continued*)

Problem	Probable Causes	Recommended Corrections
	Steam leakage at the governor valve	Check and adjust or replace parts as required.
	Restricted governor travel	Check and adjust governor so that it can move the valve to its seat.
Throttle tripping without cause (turbine not operating at excessive speeds)	Worn or defective throttle valve trip	Inspect and replace if necessary.
	Trip trigger positioned too close to the shaft	Check and adjust so that the trip trigger is properly positioned.
	Defective or broken trigger spring	Check and replace with new trigger spring.
	Improperly compressed overspeed governor spring	Check and adjust spring tension.
Unstable and unreliable governor function	Defective, worn, or sticking parts	Check, disassemble, clean, repair, and/or replace parts.
	Binding governor linkages or sticking valve stem	Inspect, disassemble, clean, lubricate, and readjust.
	Leakage at the governor valve.	Inspect for any deposits and remove. Valve may require remachining or replacement.

TABLE 7–9 (*continued*)

Problem	Probable Causes	Recommended Corrections
	Governor valve not properly seated or attached on stem	Check, align and tighten.
	Improper friction within the speed governor	Inspect, disassemble, clean, and repair or replace worn parts.
	Drains clogged at the operating valve stem bushings	Check for accumulation of any foreign material and remove from drains.
Sticking throttle valve	Dirt or foreign material in the valve	Inspect, remove material, and check strainer.
	Valve stem packing too restrictive or tight	Check and loosen packing; if problem persists, replace packing.
	Valve stem guides damaged or dirty.	Check, repair or replace.
Turbine vibration	Worn bearings	Inspect and replace.
	Misaligned couplings	Check and realign.
	Worn coupling	Check, repair or replace.
	Insufficient packing clearance	Increase clearance.
	Turbine rotor not in balance	Inspect for damaged rotor blades and rebalance rotor.
Packing leakage	Worn rings	Check and replace rings.
	Damaged or broken packing springs	Check and replace springs.

TABLE 7–9 (*continued*)

Problem	Probable Causes	Recommended Corrections
	Foreign material between ring and gland case groove	Inspect, disassemble, and clean.
Moisture in oil	Packing leakage	Check and correct.
	Cooling jacket or coil leakage	Check, repair or replace bearing bracket or cooling coils.

Troubleshooting turbines and steam traps is also an important part of plant maintenance programs. The testing of steam traps is made easier with the use of a *test stand*. A good design is one with a steam reducing valve, connection to pipe water, and compressed air in one of the feeding legs into the trap.

Two troubleshooting and diagnostic charts are presented in this section. Table 7-10 addresses power steam traps and Table 7-11 turbines.

TABLE 7–10

Troubleshooting Steam Traps

Problem	Probable Causes	Recommended Corrections
Trap not discharging	Steam pressure too high; malfunctioning pressure regulator; low boiler pressure; or steam pressure raised without adjusting trap	Check pressure readings and adjust. Consult manufacturer or trap supplier for advice on trap adjustments.
	Clogged strainer, valve, trap fitting head, and/or internal parts	Inspect and clean affected component.

TABLE 7–10 (*continued*)

Problem	Probable Causes	Recommended Corrections
	Leakage or opening at the bypass	Check; close, repair, or replace.
	Damaged or worn internal parts	Inspect, repair or replace.
Trap not shutting off	Load exceeding trap capacity	Adjust load to trap capacity rating.
	Malfunctioning or broken component	Check and repair or replace.
Trap blowing steam	Leakage in or opened bypass valve	Check; repair or replace; or close valve.
	Dirt or scale clogging trap	Disassemble and clean.
	Trap no longer has prime	Check pressure and correct.
Sudden drop in trap capacity	Low inlet or back pressure	Increase pressure.
	High back pressure	Lower pressure by opening bypass or plugged vent.
Condensate unable to drain	Air bound system caused by undercapacity vent	Install proper size vent.
	Low steam pressure	Increase pressure.
Insufficient steam heat	Defective thermostat	Check and replace.
	Boiler priming	Add freshwater to boiler while boiler blowing down.
	Vacuum pump not shutting off, caused by leaks	Inspect for cracks and repair.

TABLE 7–10 (*continued*)

Problem	Probable Causes	Recommended Corrections
	Excessive water hammer in system	Check for proper size drip-tap.
Traps freezing during cold weather	Discharge line not properly pitched	Check and correct pitch.
	Inadequate trap and pipe insulation	Insulate properly.
Back flow in return line	Incorrect fittings at the trap located below the return main	Inspect and correct according to manufacturer's recommendations.
	Discharging of high pressure traps into low pressure return	Correct piping.

TABLE 7–11

Troubleshooting Power Plant Turbines

Problem	Probable Causes	Recommended Corrections
Excessive or overspeeding	Governor improperly adjusted	Check and adjust.
	Governor valve stem and/or trip weight sticking	Adjust trigger, remove weight and refinish, and clean springs.
	Worn or broken parts causing steam to leak through the governor valve	Check, repair or replace parts.
	Improper travel of governor	Readjust governor.
Governor unstable	Dirty, sticking, or worn parts	Disassemble, clean, repair, and/or replace affected parts.

TABLE 7–11 (*continued*)

Problem	Probable Causes	Recommended Corrections
	Leakage in the governor valve	Repair or replace valve.
	Governor linkage binding or sticking valve stem	Disassemble, clean, repair, and/or replace affected parts.
	Worn governor parts	Repair or replace.
Throttle kicking in without overspeeding	Trip latch worn	Replace.
	Worn latch seats	Repair or replace latch seats.
	Damaged or broken trigger spring	Adjust spring tension or replace.
	Trip trigger located too close to the shaft	Adjust and properly position to manufacturer's recommendations.
Valve open when tripping mechanism engages	Corroded latch; incorrectly engaged or sticking valve stem	Clean and readjust.
Sticking throttle valve	Packing too tight around the valve stem	Disassemble and repack or install new packing.
	Dirt in valve or damaged by deposits	Clean, repair or replace.
Unnecessary vibration	Worn bearings	Replace.
	Misaligned couplings or shaft	Realign.
	Worn coupling	Check, repair or replace.

TABLE 7–11 (*continued*)

Problem	Probable Causes	Recommended Corrections
	Turbine rotor out of balance	Repair any blade damage and balance rotor.
Leakage in the packing	Worn rings and/or broken packing springs	Replace rings and springs.
	Accumulated foreign material between ring and gland case groove	Disassemble and clean.
Moisture in the oil	Cooling jacket damaged	Check and repair or replace bearing bracket or cooling coils.
	Packing leaking into the bearing	Repair.
Bearing fails	Poor lubrication	Correct.
	Defective oil rings	True up or replace oil rings.
	Improper supply of oil	Maintain proper oil level.
Turbine not capable of handling load	High exhaust or low inlet pressures	Check readings and maintain proper readings.
	Steam strainer dirty	Check and clean.
	Malfunctioning governor	See above.
	Nozzles and buckets clogged	Check and clean.
Thrust bearings failing	Buckets clogged	Clean.
	Improper turbine and drive unit alignment	Check and realign.
	Incorrect thrust clearance	Repair and adjust.

TRANSPORTATION EQUIPMENT

Transportation equipment—electric and gas-powered industrial trucks—is widely used for moving products and raw materials from one location to another in both commercial and industrial settings.

ELECTRIC INDUSTRIAL TRUCKS

Most electric industrial trucks are used for lifting and are ruggedly built. However, this does not mean that a regular preventive maintenance program is not needed. In fact, it is recommended that the following items be checked at the start and end of each shift:

1. Operation of steering.
2. The hydraulic system.
3. Operation during travel and hoisting speeds.
4. Operation of all lights and attachments.
5. Battery terminal connections and fluid levels.
6. Operation of seat and operating brakes.

Most maintenance programs established for electric industrial trucks are based upon length of operation—usually expressed in hours. Presented in Table 7-12 are maintenance procedures that should be followed after various lengths of truck operation.

TABLE 7–12
Maintenance of Electric Industrial Trucks

Length of Service	Maintenance Operations
100 Hours	Check differential for backlash and excessive noise.
	Check wheel bearings for fit, making sure that they are not loose.

TABLE 7–12 (*continued*)

Length of Service	Maintenance Operations
	Check tire wear.
	Check chain rollers for lubrication, and make sure that they are free running.
	Inspect chains for adjustment and anchoring. Clean and coat with light oil.
	Examine hydraulic cylinders.
	Check motor, insulation on armature, field coils, and terminals.
	Disconnect battery plug and clean out.
	Inspect commutator. If burned, clean or replace.
	Check motor coupling for proper action and lubrication.
	Examine contactors for any burning or pitting.
	Clean resistors of all dust and check insulation.
	Examine hydraulic lines for leaks.
	Check fittings and seals of hydraulic pump and valves.
	Inspect and adjust operating brakes.
	Check seat brake adjustment.
	Check seat interlock switch for proper operation.
	Be sure battery is clean.
500 Hours	Examine motor brushes for wear and proper pressure.
	Check steering gear unit for tightness. If rough or sticky, dissemble and clean.
	Check steering for equal amounts of turning in each direction.

TABLE 7–12 (*continued*)

Length of Service	Maintenance Operations
	Check all wiring for wear, chafing, burning, and loose connections.
	Inspect hydraulic system oil for any contamination. If found, drain and flush the system and replenish with clean oil.
5000 Hours	Examine lifting forks for any cracks or bending.
	Remove and dismantle upright mast checking welds.
	Clean chains with solvent; inspect and properly lubricate.
	Dismantle hoist and tilt hydraulic cylinders and clean with solvent. Inspect packing glands for wear.
	Dismantle power axle and inspect the bearings, gears, and shafts for signs of wear.
	Dismantle and clean the motor. Inspect insulation, insulator bushings, and all lead connections.
	Check all motor brushes.
	Inspect armature for proper starting and insulation.
	Examine commutator for any signs of overheating. Check connections.
	Examine armature shaft for straightness.
	Inspect field coils for proper connections, and test for insulation, shorting, grounds, and circuitry.
	Clean and inspect motor coupling.
	Check motor anchor screws for proper size and mounting.

TABLE 7–12 (*continued*)

Length of Service	Maintenance Operations
	Inspect contactors, bearings, and shafts. Check coils and shunt cable.
	Remove and clean switches.
	Clean and inspect master accelerator.
	Dismantle, clean, and inspect trailing steering axle.

GAS-POWERED INDUSTRIAL TRUCKS

Gas-powered trucks may be either diesel or gasoline powered. Like electric industrial trucks, gas-powered trucks require a detailed maintenance program. In addition, the following items should be checked at the beginning and end of each shift:

1. Brake serviceability and operation as well as the hand or brake lock.
2. The throwout bearing grease cup of the clutch or dynatork drive.
3. The operation of all gages.
4. Crankcase level.
5. Air and fuel supply.
6. Transmission oil level.
7. Tires, and head and stop lights.
8. Radiator, water level, and antifreeze.

Table 7-13 describes maintenance procedures for all gas-powered trucks, categorized according to hours of service.

TABLE 7–13

Maintenance of Gas-Powered Industrial Trucks (Gasoline and Diesel Engines)

Length of Service	Maintenance Procedure
40 Hours	Check oil level of drive axle and differential.
	Check axle air vent.
	Inspect master cylinder and lines of brake system.
	Check clutch pedal adjustment.
	Examine cooling system core and hoses, and check for leaks.
	Inspect fan and fan belt tension.
	Check battery fluid level, hydrometer reading, cables, receptacle, and terminals.
	Check generator and distributor lubrication.
	Drain engine oil and refill; check filter.
	Check fuel pump, diesel fuel oil filter, and fuel filler cap.
	Check hydraulic system pump, valve, and tank. Clean tank filler cap.
	Check lift chain adjustment, lift brackets and slides; lubricate inner slides.
	Check steering axle grease fittings, steering gear, lubrication, and mounting.
	Check transmission fluid pressure.
	Lubricate all grease fittings.
	Check linkages.
	Steam clean.
300 Hours	Check starter motor lubrication, generator, and starter motor bushes.
	Adjust points in distributor, and inspect motor, cap, and condenser.

TABLE 7–13 (*continued*)

Length of Service	Maintenance Procedure
	Check ignition timing.
	Check spark plug gap and clean.
	Inspect and adjust engine valve tappets.
	Check engine compression.
	Check cylinder head and manifold nuts, gaskets, and leaks.
	Check governor for speed and surge.
	Make carburetor adjustments.
	Check hydraulic system lift brackets and slides; tilt and lift cylinder drift; tilt cylinder glands, boots, and mounting pins; and lift cylinder packing boots and vent.
	Check steering drag link adjustment, turning radius, tie rod ends, and pins and spindle.
	Drain and refill transmission fluid.
	Drop transmission gear case and inspect.
	Inspect spring shackle U-bolts and clips and U-joints; tighten all bolts, nuts, and capscrews.
1000 Hours	Clean and repack drive axle and differential gears and bearings. Drain and refill.
	Check brake shoe lines and connections and adjust brake alignment.
	Inspect fork lift hydraulic system.
	Inspect wheel bearings of steering system.

TROUBLESHOOTING TRANSPORTATION EQUIPMENT

Troubleshooting gasoline-powered engines does not seem as mysterious as troubleshooting other equipment, because almost everyone owns an automobile or truck and is familiar with its operation and care. In contrast, diesel engine diagnosis causes more concern due to its limited use in domestic vehicles. Table 7-14 provides a general overview of common diesel- and gasoline-powered engine problems and recommended corrections.

Special note should be given to electric-powered vehicles. To avoid repetition of information, refer to Chapter 5 on maintenance and troubleshooting electrical equipment.

TABLE 7–14
Troubleshooting Diesel and Gasoline Powered Engines

Problem	Probable Causes	Recommended Corrections
Diesel Engines		
Noisy and smoking engine	Incorrect fuel-air ratio	Increase air intake into cylinders.
	Leaking head gasket	Tighten head bolts or replace gasket.
	Carbonization or fouling of valves or injectors	Clean, regrind, and reseat valves.
	Improper fuel	Check and replace fuel.
Engine knock	Improper injection timing	Reset injection timing.
	Sticking spray valve	Clean and replace worn or damaged parts.
	Bearings and/or pins exceed operating tolerances	Refit bearings and pins.
Overheating engine and loss of power	Piston binding	Remove and free up piston.
	Poor circulation of cooling system	Clean and refill.

TABLE 7–14 (*continued*)

Problem	Probable Causes	Recommended Corrections
	Improperly timed engine injection	Reset injection timing sequence.
	Inadequate lubrication	Provide for adequate lubrication. Clean system.
Erratic engine starts	Leaky gasket	See above.
	Valves not properly functioning	Clean, regrind, and reseat valves.
	Fuel pump malfunctioning	Clean filter, vent air binding, or clean out pump.
	Water in fuel	Drain fuel tank and run off water.
Poor engine performance (carbon buildup)	Using wrong fuel	Check and change fuel.
	Sticking fuel injector plungers	Clean.
	High back pressure on exhaust	Check and clean exhaust system.
Gasoline Engines		
Difficult engine start	Contaminated fuel	Check, clean line and replenish with new source.
	Too rich or too lean a gas-air mixture	Adjust carburetor, check choke.
	Timing off	Retime as required.
	Gasket leaking in air	Tighten or replace head and manifold gaskets.
	Insufficient voltage supply	Check battery, connections, and spark transformer.

TABLE 7–14 (*continued*)

Problem	Probable Causes	Recommended Corrections
Engine missing	Defective spark plug	Clean or replace.
	Leaky gaskets	See above.
	Worn or damaged valves	Repair valves and grind seats.
Engine knock	Low octane fuel	Check and increase octane fuel rating.
	Carbon deposits loaded in cylinders	Check and clean.
	Loose and/or worn connecting rod bearings	Replace.
	Worn main crankshaft bearings	Replace.
	Worn cylinders	Refit with larger sized piston, or bore if out of true.
	Improperly timed engine	Reset timing.
Engine overheat	Cooling system ineffective	Clean out system and replenish with fresh antifreeze.
	Improper lubrication	Drain and flush oil refilling with clean and correct grade oil.
	Loose fan belt	Correct belt tension.
Inconsistent engine performance	Air leak in fuel line	Check and correct.
	Exhaust valve leak	Check and correct machine seating or replace valve.

TABLE 7–14 (*continued*)

Problem	Probable Causes	Recommended Corrections
	Worn rings	Replace rings.
	Cylinder head gasket leaks	Tighten or replace.
	Improper timing	See above.
	Dirty carburetor	Clean.

VENTILATING EXHAUST SYSTEMS

The maintenance of exhaust systems primarily involves moving components—that is, fans and fan motors. The major concerns in ventilating system maintenance are keeping the system clean and free from corrosive and flammable fumes and abrading dust. Periodic cleaning, especially of fans, blower blades, and housings, is essential. Buildup of foreign material can dramatically cut down on the performance and efficiency of the unit.

FANS

An important element of the entire ventilating system, fans enable air movement. There are three basic fan designs: propeller, axial, and centrifugal.

1. In *propeller designs* fans that look like propellers are mounted in a ring or panel. Air flow is in a straight line and parallel to the shaft. Propeller fans are primarily used for very low pressure-producing conditions.

2. In *axial designs* a propeller fan is enclosed in a cylindrical housing that is available in a variety of air moving capacities. Axial fans are used at static pressures up to 4 inches.
3. In *centrifugal designs* a wheel is enclosed in a scroll-shaped housing. Air enters through the eye of the rotor parallel to the fan shaft and discharges through the housing outlet that is 90° to the shaft.

Fan selection should be based on several important factors, including the quantity of air to be moved, required fan static pressure, unit cost, available space, noise parameters, types of elements in the air, and operating temperatures. Table 7-15 presents basic maintenance procedures recommended for fan units.

TABLE 7–15
Maintenance of Fan Units

Fan Component	Maintenance Procedure
Foundations	Check for rigidity and proper anchoring.
	Tighten any loose fasteners.
Base	Base should be supported by vibration isolator: check for sufficient number.
	Check that connections are flexible between fan and ductwork.
Bearings	Check that bearings, initially lubricated by manufacturer, are filled to proper level before placing in operation.
	Check for proper alignment.
Couplings	Check to make sure that proper alignment is maintained.
	Lubricate according to manufacturer's specifications.
V-belt drives	Check for proper installation and alignment.
	Make sure that belt is properly mounted and has correct amount of tension.

When a new fan is installed, it should be carefully observed for the first couple of hours. After a few days of operation, one should check:

1. Amount of vibration (it should not be excessive)
2. Lubrication level and leakage
3. Operating temperature of bearings
4. Tightness of all screws, bolts, and riveted joints
5. Belt drive tension
6. General condition of the unit.

MOTORS

Totally enclosed motor units must be free from accumulations of shop dirt such as paint, sawdust, and lint. Excessive buildup of dirt will prevent the motor from cooling properly, which will lead to overheating and eventual failure.

ROOF VENTILATORS

Most roof ventilators are made of galvanized sheet metal and should be allowed to weather for several weeks before being painted. Ventilators should be inspected periodically for signs of rusting, debris buildup, and inadequate flashing. Drain and weep holes should be checked to see that they are not clogged. This can be accomplished with a semiannual inspection.

PART THREE

ABCs OF THE TRADE

8

The ABCs of the Trade

Every trade requires a basic body of knowledge and skills applicable to most jobs within it. This chapter presents basic information important to the machine and maintenance mechanic, whether he or she is concerned with building, electrical equipment, mechanical equipment, or service equipment maintenance and repair. Operational procedures pertaining to many different types of machine and maintenance mechanic jobs are then discussed in a step-by-step format that is easy to follow.

BASIC INFORMATION OF THE TRADE

This section reviews three major topics—corrosion, lubricants and lubrication, and welding—that all machinists and maintenance mechanics should be knowledgeable about regardless of their specific jobs.

CORROSION

Corrosion is one of the most important and common causes of machine and equipment deterioration and failure. The importance—destructiveness—of corrosion was highlighted in a National Bureau of Standards study that found that over a twenty year period corrosion contributed to an annual economic loss in excess of $10 billion per year, or more than 2% of the U.S. total Gross National Product (GNP).

Corrosion can be defined as destructive chemical or electrochemical attack on materials. Direct chemical attack occurs only to materials exposed to corrosive atmospheres, solutions, and/or temperatures. An example of this is the formation of iron oxide (rust) in overheated boiler tubes that is caused by metal contact with a strong acid or alkali. Most metal corrosion, however, is a result of an electrochemical reaction that occurs in two parts. The first is the anodic reaction, in which the metal dissolves in a solution in the form of positive ions. The second is the cathodic reaction, in which the positive ions coat the surface of the negative (cathode) material. Magnesium (and its alloys), aluminum, manganese, zinc, and iron typically corrode at the anodic end of the reaction; gold, lead, silver, and copper are cathode end materials.

Corrosion control begins with the proper selection, design, fabrication, installation, and maintenance of products. Selection of product materials should be based on past experiences—how well that material functioned in other similar situations. If this is not possible, review data pertaining to the chemical resistance and limitations of the particular material.

TYPES OF CORROSION

The machine and maintenance mechanic needs to be able to identify the different types of corrosion, so that proper prevention procedures and/or corrective actions can be taken.

General Corrosion. General corrosion is the uniform thinning and loss of metal. Of all forms of corrosion, this type causes the greatest amount of damage. Practices that may prevent or slow down corrosion include the use of inhibitors, or protective coatings or linings, and environmental changes such as the reduction of operational temperatures, acidity, aeration, and velocity of material flow. When extensive corrosion is present, it may be necessary to change to a metal that is more resistant.

Pitting. Pitting is electrochemical corrosion at anodic areas. Although pitting is localized, with minimal general corrosion, it is the most serious form of corrosion because it will rapidly destroy the metal.

Pitting can occur in any metal, but it is most common in aluminum and stainless steels that are exposed to water containing metal chlorides. In most cases, pitting can be minimized by using an inhibitor, such as sodium or potassium dichromate.

Galvanic Corrosion. Galvanic corrosion occurs when two dissimilar metals are brought into contact and the less noble metal—that is, the metal with more oxide—is attacked. The severity of corrosion depends on both the degree of potential difference between the metals and the amount of surface contact.

Most galvanic corrosion is localized, occurring in places such as where bronze valves and fittings are used in combination with steel piping. In some cases, galvanic corrosion can be used to advantage, such as the use of aluminum and zinc anodes to protect steel and iron pipelines. The most common way to correct galvanic corrosion is to replace one of the metals with one having a more similar noble characteristic.

Dezincification. Dezincification occurs in brass alloys (a mix of copper and zinc) containing more than 15 percent zinc. In the corrosion process the zinc is selectively removed, leaving a "sponge copper." Dezincification can be prevented by selecting brass products that are manufactured with inhibitors such as arsenic, antimony, or tellurium.

Graphitization. Graphitization is the type of galvanic corrosion that occurs in cast irons. In these products, the graphite and iron constituents react to greatly reduce the strength of the product. This form of corrosion is common in buried cast iron lines exposed to moist, acidic conditions.

When graphitization occurs, the product will usually have to be replaced with a less reactive metal.

Oxidation. Oxidation often identified in the form of metal scaling, generally occurs at elevated temperatures and results in the formation of metal oxides. When this occurs, there will be a loss of metal. Corrective procedures include the reduction of operational temperatures or the replacement of the metal with alloys containing a minimum of 25 percent chromium and 20 percent nickel.

Sulfidation. Sulfidation occurs when sulfur-bearing atmospheres attack a metal at elevated temperatures, in most cases, at temperatures exceeding 600°F. This type of corrosion is primarily limited to nickel and nickel-bearing alloys. Sulfidation can be minimized by reducing operational temperatures or by selecting nickel alloyed products that include chromium.

USE OF NONMETALLIC SUBSTITUTES

Many nonmetallic products have proved to be good substitutes for metals that have recurring corrosion problems. The substitutes include polymers, elastomers, brick, glass, and wood. To cover all the chemical and corrosive properties of these materials far exceeds the intent of this book. However, a brief description of each is provided to give the apprentice and mechanic some idea of how these materials perform.

Plastics. Plastics are a successful substitute for many metal products, especially in situations involving low temperature and pressure service. Several major families of plastics, including polyethylene, polyvinyl chlorides (PVC), styrene, polyesters, epoxy resins, and phenolic and furan resins, are resistant to corrosive conditions. Chemical-resistant plastics, such as polyethylene and tetrafluorethylene (Teflon), are also practically unaffected by acids and alkalis. Today, PVC is widely used for piping material and ventilating ductwork, especially in situations in which corrosive materials and fumes are handled.

Elastomers and Rubber. Synthetic rubbers and elastomers have been developed to withstand corrosion by numerous chemicals, and natural rubber resists hydrochloric and phosphoric acids. However, both elastomers and rubber are readily attacked by oxidizing acids, such as nitric and chromic solvents.

Glass. Glass is often used as a lining in commercial processing and materials handling equipment. Generally, glass is resistant to most acids and other chemicals, except hydrofluoric acid and some phosphoric acids. Field maintenance and repair of glass-lined equipment can be accomplished with plugs and cover plates of tantalum, resin cements, and Teflon.

Brick. Brick is a popular material for lining construction, used for a number of different conditions that are extremely corrosive to high alloy metals. Carbon bricks are used to withstand alkaline and acidic conditions.

Wood. Wood is still used in the construction of tanks, vats, and filter presses. It is resistant to acetic acid, but will deteriorate when exposed to oxidizing acids and strong alkaline solutions.

LUBRICANTS AND LUBRICATION

Lubrication modifies frictional properties and reduces damage and wear at the surface of moving parts. Any substance introduced between moving parts to accomplish this is known as a lubricant.

TYPES OF LUBRICANTS

Of primary concern to maintenance and machine mechanics are liquid, grease, and solid lubricants.

Liquid Lubricants. Any liquid (e.g., water or alcohol) can be used as a lubricant. However, the most common lubricants are petroleum-based or manufactured synthetic fluids. Petroleum lubricants are popular because of their relative low cost, availability, and general suitability.

Lubricant performance is determined by its characteristics or properties. These properties are determined by tests conducted by manufacturers. The tests are standardized by the American Standards and Testing Materials (ASTM) organization, providing uniformity in the field. Examples of physical properties tested for are viscosity, density, pour, flash point, and demulsibility. Chemical properties tested for include carbon residue, acidity, oxidation, corrosion, extreme pressure, and precipitation number.

Properties of Liquid Lubricants. Of all the properties, *viscosity* is considered the most important. Viscosity is the fluidity of fluids, expressed as resistance to flow. It is affected by temperature, pressure, and fluid motion. The primary instrument used to determine the viscosity of lubricants is the Saybolt Standard Universal viscometer which expresses this property in Saybolt Universal Seconds (SUS). However, the most recognizable viscosity rating system for lubricants is the SAE viscosity number, established by the Society of Automotive Engineers (SAE).

TABLE 8–1

SAE Numbers of Liquid Lubricants

SAE Viscosity Number	Viscosity Range Saybolt Universal Seconds			
	0 min.	0 max. (temperatures in degrees Fahrenheit)	210 min.	210 max.
Crankcase Oils				
5W	...	4,000	...	...
10W	6,000	<12,000	...	...
20W	12,000	48,000	...	...
20	...	...	45	<58
30	...	...	58	<70
40	...	...	70	<85
50	...	...	85	110
Transmission and Axle Lubricants				
	...	...	40	...
75W	...	...	49	...
80W	...	...	63	...
85W				
	...	15,000	...	...
75	15,000	100,000	...	...
80	...	...	75	120
90	...	...	120	200
140	...	...	200	...
250				

The temperature at which an oil will just begin to flow is known as its *pour point*. The *cloud point* is the temperature at which there is a separation of wax or hydrocarbon components. Steam cylinder oils and black oils are frequently tested for their *precipitation number,* a check for carbonaceous matter and formed asphaltenes.

The *acid and base number* is important for corrosion-resistant situations. For example, a lubricant with an acid number of 1.2 may not corrode a particular bearing metal, but one with an acid number of 0.1 would produce significant corrosion in the same metal. The *foam characteristics* of lubricants are important in crankcase, turbine, and circulating oils where foaming is a problem. An oil with small amounts of an antifoam additive should be considered for such situations.

An important property to consider where there are high pressures and rubbing velocities is *extreme-pressure* (EP). Under these conditions, the use of normal oils will lead to excessive material deterioration and corrosion. EP lubricants, on the other hand, will help prevent metal welding, scuffing, and tearing during high pressures and contact.

Greases. Lubricating greases are considered fluid lubricants, of thicker consistency. Most greases are thickened with the addition of soap, clays, and/or pigments. An important property of greases is the *dropping point,* the temperature at which the grease changes from its thickened state to that of a liquid.

Greases are classified according to a consistency number, which is established by the National Lubricating Grease Institute (NLGI). Table 8-2 summarizes this numbering system.

TABLE 8–2
Grease Consistency Numbers

Consistency Number	Appearance	Work Penetration
000	semifluid	445–475
00	semifluid	400–430
0	semifluid	355–385
1	soft	310–340
2	medium	265–295
3	medium hard	220–250
4	hard	175–205
5	very hard	130–160
6	block-type	85–115

Solid Lubricants. As its name implies, a solid lubricant is made up of one or more solids that are used to modify friction and wear. Within recent years, solid lubricants have gained wider acceptance. The basic type of solid lubricant is *unbonded* (loose), which is available in either granular or powder form. Examples of unbonded solid lubricants are graphite, molybdenum, disulfide, polytetrafluoroethylene (PTFE), polychlorotrifluoroethylene (PCFE), and talc. These lubricants can be applied by brushing, dipping, or spraying.

The second type of solid lubricant is *bonded.* Binders or adhesives are mixed with the solid lubricant and applied to the protected surface. Inorganic adhesives are recommended when temperatures exceed 1200°F (260°C). Typical bonded solid lubricants are ceramic binders with stable powered metals, such as molybdenum disulfide and graphite.

USES OF LUBRICATING OILS

There are a number of different methods used to select the appropriate lubricant. One of the most useful is by application. A general classification of lubricating oils, based on viscosity and use, is summarized as follows:

1. Low viscosity oils: 60-200 at 100°F, SUS
   a. spindle oils
   b. light machine oils
   c. light turbine oils

2. Medium viscosity oils: 220-420 at 100°F, SUS
   a. medium bodied machine oils
   b. compressor oils
   c. medium turbine oils

3. Heavy viscosity oils: 420-1500 at 100°F, SUS
   a. heavy machine oils
   b. gear oils

4. Extra heavy viscosity oils: 1500-3000 at 100°F, SUS
   a. extra heavy machine oils
   b. heavy gear oils
   c. cylinder oils

For a more detailed description, see Table 8-3, which summarizes oil classifications according to type of service.

TABLE 8–3
Classification of Lubricating Oils According to Service

Classification	Service	Viscosity SUS @ 100°F (typical)
I	Spindle oils:	
	speeds >3600 rpm	35–100
	speeds <3600 rpm	100–150
	speeds 300–1800 rpm	150–900
II	Turbine oils:	
	direct connected	150
	geared	300
	marine-type, bleeder	400
III	Hydraulic oils:	
	vane pumps	150–300
	angle and radial piston pumps	150–900
	axial piston	150–300
	high pressure, high output pumps	300
	gear pumps	150–600
IV	Circulating bearing oils:	
	light loads	150–400
	intermediate loads	400–900
	heavy loads	900–2500
	extra heavy loads	1500–3500
V	Enclosed gear oils splash-type:	
	spur, herringbone, and bevel	600–1800
	heavy or shock loading	1500–2500
	worm gears	2500
	hypoid gears	2500

TABLE 8–3 (*continued*)

Classification	Service	Viscosity SUS @ 100°F (typical)
VI	Engine motor oils:	
	gasoline engines	SAE 30
	diesel engines	SAE 30
	gas engines	SAE 30, 40
	propane engines	SAE 30
	dual trifuel engines	SAE 30, 40
VII	Way oils:	
	light loads	150–300
	intermediate loads	300–500
	heavy loads	900
	combustion way and hydraulic oil	150–300
VIII	Steam cylinder oils:	
	wet steam	2500
	dry steam to 500°F superheat	3000–4000
	dry steam over 500°F superheat	4000–6000
IX	Compressor cylinder oils other than refrigeration:	
	air and inert gases to 150 psi	300
	air and inert gases 150-1000 psi	500–600
	air and inert gases 1000-2000 psi	1500
	air and inert gases 2000–4000 psi	1500–2500
	air and inert gases >4000 psi	2500
	hydrocarbon gases	1600–1700
	pneumatic gases	200–500
	oxygen	water
	expansion engines, cryogenic <−300°F	no lubrication

TABLE 8–3 (*continued*)

Classification	Service	Viscosity SUS @ 100°F (typical)
X	Refrigerating oils with any refrigerant except SO_2:	
	reciprocating and rotary compressors with: evap. temperature >−50°F	200–300
	evap. temperature <−50°F	150
	centrifugal compressors (bearings)	300
	hermetric reciprocating compressors with: evap. temperature −50°F min.	300
	evap. temperature −40 to −70°F	150
	Sulfur dioxide refrigerant (white oil)	150–300
XI	Machine oils: light loads, high speeds	35–150
	medium loads, medium speeds	150–600
	heavy loads, slow speeds	600–1800
	very heavy loads, slow speeds	to 4000
XII	Automotive transmission oils: conventional transmission and differential	SAE 90–125
	hypoid gears	SAE 90–140
	automotive fluid transmission	200
XIII	General purpose oils: light duty service	150–300
	intermediate service	400–900
	heavy duty service	1500–2500

TABLE 8–3 (*continued*)

Classification	Service	Viscosity SUS @ 100°F (typical)
XIV	Air line lubricating oils: small pneumatic tool	200
	pneumatic control cylinders	500
	large air cylinders	800
XV	Vacuum pump oils: 0.05 microns Hg	300–400
XVI	Open gear fluids: hand, spray, or drop feed systems without diluent	400–1500 @ 210°F
	with diluent	>1500 @ 210°F
XVII	Black oils: circulation systems	400–2000
	swab, pour and dip systems	2000–5000

USES OF GREASES

Greases are generally classified by the type of thickener used during manufacturing. They may also be grouped according to chemical and physical properties, composition, manufacturing method (e.g., synthetic), and type of base. Synthetic greases are grouped by the type of synthetic fluid used during their manufacture. Table 8-4 presents a classification of common greases.

TABLE 8–4
Classification and Uses of Greases

Base (soap)	Temp. (°F) Limits	Applications
Aluminum	180–190	Water repellent, can withstand slick loads, and will not separate under centrifugal force. Used in auto chassis, flexible couplings, cams, small open gears, and eccentrics.
Calcium (lime)	175	Water repellent, buttery in texture. Used as cup grease, for plain bearings, and antifriction bearings under moderate stress and speeds.
Calcium complex	>300	Water repellent with good pumpability properties. Used for general purpose lubrication.
Graphite grease	175	Water repellent, used for wet conditions. Not recommended for high temperature or antifriction bearing use. Provides good service in rotating water screen chains, pump plungers, and hydraulic rams.
Lime-lead mixture	200	Water repellent, good pumpability, and smooth texture. Used in central grease systems with long lines, and offers good service under heavy pressures and severe operating conditions.
Lime-lithium mixture	250	Used with antifriction bearings exposed to high and/or low temperatures.
Lime-rosin cold set	180	Water repellent with a buttery consistency. Used on open gears, axle bearings, and rough machining.
Lithium	100–300	Almost water repellent, with low pour point, and good low temperature properties. Good general purpose service.

TABLE 8–4 (*continued*)

Base (soap)	Temp. (°F) Limits	Applications
Nonsoap	250–500	Insoluble in water, with a high temperature service range. Used as rust inhibitor and a good acid resistor.
Residuum	200	Will dissolve away when exposed to water; may use additives to improve water repellency. Used primarily for lubricating large open gears.
Sodium (block)	to 450	Forms emulsion with water; is hard and brittle. Used for high temperature plain journal bearings, and large slow turning journals.
Sodium (soda)	200–350	Emulsifies with water; is smooth and fibrous in character. Used with antifriction bearings at high and moderate speeds.

WELDING

An important process associated with the machine and maintenance mechanic trade is the joining of two or more base metal parts by brazing or welding. In fact, the original use of welding and brazing was for the repair of damaged parts.

Welding is a process whereby two metals are melted in such a way that their liquids combine and harden or fuse together. In many cases, a *filler metal,* such as a welding rod, is used to facilitate the process.

Brazing uses a principle known as *capillary attraction* to join metals. The filler—brazing rod—is melted so that it does not lay flat on the base metal, but bulges and floats through capillary attraction into the joining surfaces. The brazing metal does not become a liquid solution with the base metal. Technically, brazing is accomplished at temperatures above 800°F (427°C); joining metals at temperatures under 800°F is known as soldering.

Within the welding trade, welds may be indicated in technical drawings. Manufacturers who recommend welds give clear symbolic representations for the specific type of weld to be used. Each symbolic representation includes an elementary symbol which may be completed by a supplementary symbol, dimensional specifications, or some complementary indications (especially for shop work). (See Figure 8-1.)

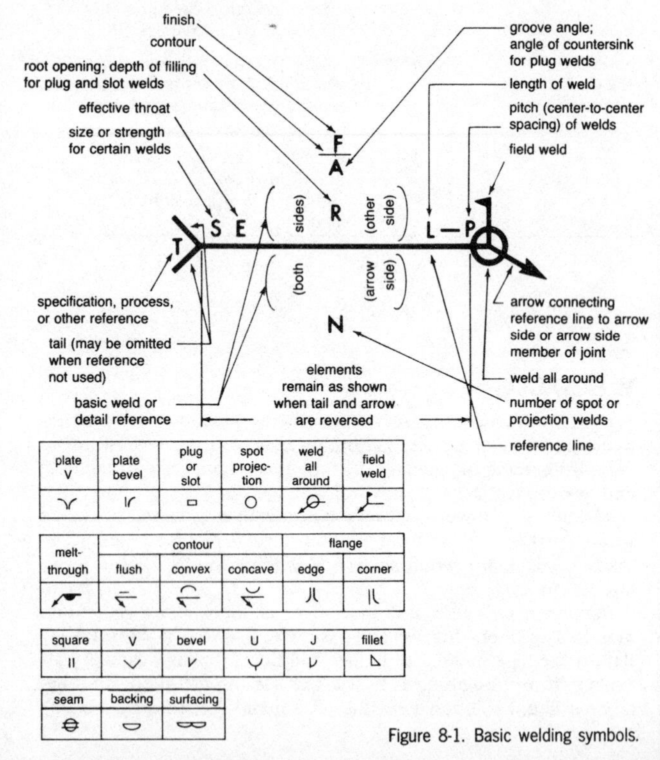

Figure 8-1. Basic welding symbols.

GAS WELDING

Gas welding is accomplished by burning a gaseous mixture that consists of oxygen and another gas (or gases). It is commonly used for repair work, but is being replaced by other joining processes in manufacturing processes.

Types of Gases Used

1. *Combustible gases* are sometimes referred to as fuel gases, because they supply the fuel for the flame. There are four major industrial combustible gases: acetylene, natural gas, propane, and methylacetylene propadene; of these, acetylene is the most commonly used.
2. *Oxidants* serve to promote the rapid burning of the gas mixture and are made up of pure oxygen or gases that have a large percentage of oxygen. In most maintenance shops pure oxygen is used.
3. *Inert Shielding Gases* are used to exclude oxygen and nitrogen from the welding area. They are primarily used in arc welding operations, though they are sometimes incorporated in gas welding procedures.

Type of Flame Used. The type of flame used in welding depends on the type of weld desired and the metals being joined. Each type of flame is obtained by adjusting the proportion of oxygen to fuel gas mixture. This mixture can be accomplished in two ways. The first, known as *nozzle mix,* is limited to industrial furnaces. In this method the oxygen and fuel gas are mixed at the outlet of the torch. The second method is *premix,* in which the gases are mixed before they leave the torch. Premix is used for the vast majority of welding operations.

The three types of flames employed for welding are carburizing, neutral, and oxidizing. A carburizing flame has more fuel than oxidant gas. A neutral flame has an equal mixture of oxidant and fuel gases. An oxidizing flame has a higher proportion of oxidant. Table 8-5 gives the specifications for each type of flame.

TABLE 8–5
Welding Flame Specifications

Flame Type	Oxygen to Acetylene Ratio	Max. Flame Temp. (degrees F)
Carburizing	0.8:1	5550
	0.9:1	5700
Neutral	1:1	5850
Oxidizing	1.5:1	6200
	2:1	6100
	2.5:1	6000

Methods of Gas Welding. Although gas welding can be accomplished by automatic procedures, the vast majority of gas welding is done manually. Two basic types of manual welding techniques differ in the relative position of the welding torch to the welding rod and weld path.

In *forehand welding* the welding rod precedes the flame, while the torch is pointed in the direction of the weld. Complete heat distribution and metal flow is accomplished by moving the torch tip in a small circular motion. This technique is recommended for welding either ferrous or nonferrous metals.

Backhand welding is the opposite of forehand welding. Here, the torch is pointed in a direction opposite to the weld path—backwards. The rod is placed between the flame and the molten pool of metal. Again, heat distribution and metal flow is assured by a circular movement of the torch. The backhand technique is frequently used in pipe and vertical welding where the solidified weld is to serve as a step or support for the molten pool of metal.

Special Operation—Flame Cutting. Another operation that can be accomplished with gas welding is flame cutting. Also known as *oxygen cutting,* this procedure employs a high velocity oxygen and gas flame to melt and remove metal. This procedure requires the use of a *cutting torch*, which resembles a welding torch. However, the tip of the cutting torch is made up of a central hole surrounded by a ring of outer holes. The center hole is used to provide the oxygen-acetylene mixture for the flame (as in a regular welding torch), while the outer ring supplies the high velocity oxygen.

TABLE 8–6
Welding Technique Considerations

Metal	Recommended Technique	Flame Type	Comments
Mild steel	Forehand or backhand	Neutral	Backhand method will produce less distortions in the metal.
Gray cast iron	Backhand	Neutral	Filler rod used should be either cast iron or nickel.
Copper-zinc brasses	Forehand or backhand	Oxidizing	Fluxes should be added to the welding rod.
Copper	Forehand or backhand	Neutral or carburizing	Deoxidized filler rods should be used.
Aluminum	Forehand or backhand	Slightly carburizing	Use minimum fluxes during welding and clean after weld.
Stainless steels	Forehand or backhand	Neutral	Filler rod should have 1–1.5 percent chromium content.

In metal cutting, the metal is heated to temperatures slightly below the molten state. At this time, high velocity oxygen is applied that rapidly melts and removes the metal. Flame cutting is used for a variety of maintenance and repair work including trimming, gouging, and edge preparation.

ARC WELDING

The term "arc welding" is used to describe a variety of welding processes, such as shielded, submerged, gas-tungsten, arcspot, and gas-metal arc welding. In each case, an electric arc(s) is established between the workpiece and a welding electrode. Either direct or alternating current may be used. Arc welding can be either manual, partially automatic, or fully automatic. However, most maintenance and repair work makes use of manual units.

The welding arc consists of four components: an electrical field, a magnetic field, ionized gas, and molten metal. Because both AC and DC power supplies can be used, care must be taken in their selection and use. If an AC arc is desired, both positive and negative terminals will be capable of supplying equal amounts of heat. A DC arc, however, only provides heat along the positive (anode) terminal.

Arc welding units are available in AC or DC designs. Both types may come with either motor generators, transformers, or driven generators. The first two types have a primary lead that connects to a main power supply. Driven generators, by comparison, can be operated by either electric motors or gas engines.

Electrodes. Critical to arc welding is the proper selection of electrodes, which serve the same function as the gas welding rod. Welding electrodes come in a wide variety of specifications. To maintain consistency, the American Welding Society (AWS) has developed a coding procedure, known as the AWS Number Code, that consists of a series of alpha-numeric symbols describing electrode characteristics. Let's use an AWS Number Code of E6011 marked on an electrode to explain the code. The "E" notes that the rod is to be used for electric arc welding. The next two digits, "60," indicate the rod's minimum tensile strength in 10,000 lbs. psi. (In our example, the rod's minimum tensile strength is 60,000 pounds [lb] per square inch [psi].) The third digit, "1," tells the welding position that the rod should be used in. (1 indicates all positions, 2 horizontal and flat position, and 3 flat position.) The last digit, which can range from 0 to 6, is used to describe the specific welding characteristics of the electrode. (See Table 8-7.)

Many electrodes are available with a flux coating and are referred to as *coated electrodes.* The basic types of flux coatings are cellulose, rutile, rutile and basic compounds, iron oxides, manganese-iron-silicates, and calcium carbonates. In addition to flux coatings, electrodes are also classified according to specific welding characteristics: fast-freeze, fill-freeze, low-hydrogen, iron powder, and 700xx through 100xx series electrodes. To select the proper flux coating and welding characteristics, the mechanic should refer to manufacturer's recommendations and AWS standards.

TABLE 8–7
Fourth Digit Electrode Characteristics

Fourth Digit	Welding Bead Produced	Polarity	Weld Quality	Other Comments
0	Flat or concave with deep penetration	DC-reverse	High	AC or DC may be used if last 2 digits are 20 or 30.
1	Flat or slightly concave	DC-reverse or AC	High	
2	Convex with fair penetration	DC-straight or AC	Medium	Medium-sized arc.
3	Slightly convex shallow penetration	DC-reverse or straight AC	Medium to High	Soft arc.
4	Medium penetration	DC-reverse or straight AC	. . .	Use for deep groove welds; has easy slag removal.
5	Flat to slightly convex with moderate penetration	DC-reverse	High	Soft arc.
6	Flat to slightly convex with moderate penetration	AC	High	Soft arc.

BUILDING AND PLANT MAINTENANCE PROCEDURES

Maintenance and repair procedures for building and plant facilities are important to all mechanics. Of primary concern are both the correct application of techniques and materials and the most efficient and effective sequencing of operations. This section discusses how basic jobs should be accomplished in this area.

CONCRETE FLOORS

Considered the most important part of any industrial or commercial building, floor surfaces should be serviceable: able to handle materials easily and easily kept clean.

Generally, concrete floors should not be difficult to maintain. However, within industrial and commercial buildings, such a flooring is considered troublesome. Lack of good workmanship and ignorance of the basic principles of installing concrete floors are the major reasons for this.

The strength of concrete is determined by the amount of mixing water used in the batch, which is expressed as a ratio of water to volume of cement. As long as the concrete is workable and the aggregates are clean and structurally sound, the material will be strong. The strength of the concrete decreases as the water ratio increases. Unfortunately, some workers misinterpret this and conclude that high-strength concrete is based solely on a low water-to-cement ratio, and use too little water.

PREPARATION OF CONCRETE FLOORS

There are two methods used to finish concrete floors: monolithic finishing and deferred topping finishing.

Monolithic Floors. Monolithic floors are made up entirely of ready-mix concrete that has too much water in it to allow it to reach its maximum strength. Such surfaces, without anything added, will be weak and porous. Monolithic finished floors will become dirty and rough unless they are subjected to only very light use.

There are two techniques that can be used to strengthen monolithic floors. A description of each procedure follows.

Densification of Floors. Monolithic floors can be densified with the addition of aggregate material. The best way to do this is to mix into the surface of the laid concrete a large amount of tough, hard, aggregate material. The material should be well mixed with cement with a low water cement ratio—about 1 1/2 to 2 pounds per square foot. Properly installed, such floors can handle moderate-to-heavy use if the base slab is at least 2 or 3 inches thick.

Use of Floor Hardeners. There are liquid floor hardeners, such as chemicals that react with the free lime, and sealants that hold down the dust caused by floor disintegration. Waxes, oils, varnishes, and common sweeping compounds also have a similar effect.

Deferred Concrete Floors. Applying a concrete floor to a slab that has already been set produces a surface that is very strong and tough. In a deferred concrete floor, the water-to-cement ratio is carefully controlled and greater amounts of fine and coarse aggregates are used. Of the two finishing methods, this method is preferred. If properly installed, deferred concrete floors should never require sealing.

Installing Deferred Concrete Floors

Step 1 *Prepare subsurface of concrete.* For new construction, the base slab should be 3/4 inch below the finished grade. When partially hardened, the concrete should be brushed with steel wire to remove all laitance and scum, thereby exposing a clean and rough coarse aggregate surface.

Step 2 *Clean prior to installation of topping.* Since the subsurface is often laid days or weeks prior to topping, it must be thoroughly cleaned of all dirt, oil, grease, concrete, paint, and other foreign substances that may have accumulated on it. One day prior to topping, the slab must be saturated with water to prevent it from drawing too much water from the topping during the curing period.

Step 3 *Apply cement grout.* A coating of cement grout should be applied after the subsurface has been saturated with water. The grout acts as an adhesive or sticking agent to hold the topping to the base slab.

Step 4 *Batching of the mix.* Because the stone aggregate is the strongest part of a concrete floor, the best flooring will have as much aggregate as possible. Therefore, a mixture of 1 cubic yard of topping mix should have 1 full cubic yard of coarse aggregate. The aggregate used is usually a basaltic or granitic rock that has been tested for hardness. Approximately 5 gallons of water per cement sack is used. (About 12 sacks of cement will produce 1 cubic yard of concrete.)

Step 5 *Removal of water.* Once the concrete topping has been laid or *screeded*, the water used for workability must be removed. This can be accomplished by laying burlap cloth on top of the mix.

Step 6 *Troweling.* Troweling is the final step, often done by both machine and hand methods. When the watery film on the surface has evaporated, machine troweling is used. Usually, several passes with the trowel are necessary. Hand troweling is the finishing step. In this procedure downward pressure must be applied at the proper blade angle. Hand troweling is extremely important because it helps to tighten the surface. As the rate of evaporation will vary, so does the troweling rate. Often a number of trowelings will be required to obtain a hard, dense floor.

Step 7 *Curing.* One of the most overlooked steps in the treatment of a floor after troweling is curing. The concrete should be kept moist so that the cement will continue to combine chemically with the water. The longer that the concrete can remain moist, the stronger, more impervious, and wear-resistant it will become. Concrete floors should not be used for 5 days after pouring. It is recommended that only light activity be conducted on the floor for an additional 10 days.

ATTACHING EQUIPMENT TO FLOORS

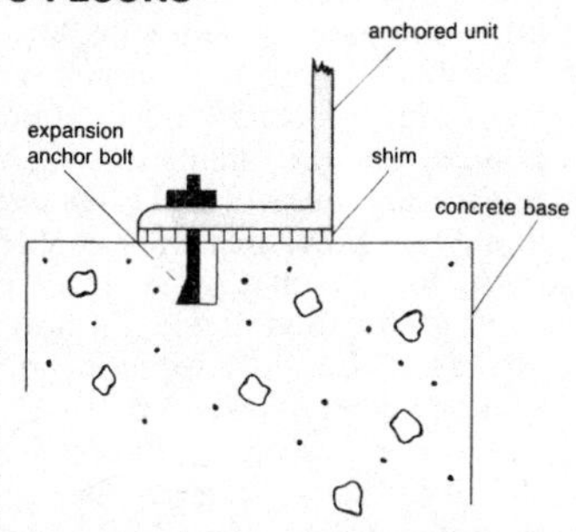

Figure 8-2. Typical expansion anchor bolt detail.

In many industrial and commercial facilities, it is necessary to fasten equipment and other machinery to concrete floors. Expansion anchors and bolts are commonly used to do this. (See Figure 8-2.)

The concrete must be able to withstand the stresses created by the equipment to be attached. When large bolts are used, the base course should be made of a concrete that is well proportioned with a maximum of 6 gallons of water per sack of cement. This mixture will provide for a good grade of concrete. After the floor has hardened and cured, mark the positions for the bolts on the floor. Then drill holes to the depth required for insertion of the expansion anchor.

REPAIR AND TREATMENT OF DEFECTIVE CONCRETE FLOORS

There are a number of problems commonly encountered with concrete floors. Some are the result of poor construction, while others are due to natural usage. Appropriate treatment minimizes damage caused by normal servicing, and with proper repair, defects can be corrected.

Dusting. The formation of concrete floor dust can be minimized by one of the hardening treatments discussed above (e.g., monolithic finishing, chemical hardening, and waxing). A thin layer of dust caused by normal working loads can be removed by attaching steel wool pads to a scrubbing machine and thoroughly cleaning the floor. In more severe cases, the surface may require grinding before treatment. Once the dust is removed, a sealant can be applied.

Cracking. Cracks in concrete floors can be structural, originating in the base and extending through the surface, or superficial, occurring as a result of normal service.

Wearing, or *crazing,* cracks, if not too deep, may be removed by grinding and then filling the cracks with varnish or resin. Though the crazing will still be visible after filling, accumulated dirt and liquids are prevented from entering the damaged area. Synthetic resins (e.g., Cumar®) should be powdered and dissolved in the appropriate solvent (e.g., xylol) in an approximate ratio of 6 pounds of resin to 1 gallon of solvent. For wider cracks, cement is often added to the resin.

When cracking is severe and/or the cracks are deep, the area should be replaced. The damaged area should be chipped off to a depth of at least 1 inch, roughened, and thoroughly cleaned. Then, the chipped surface and surrounding area should be saturated with water for several hours before new concrete is placed in the area. Figure 8-3 shows correct and incorrect methods of patching, correct and incorrect methods of screeding a patch, and how a patch can be protected.

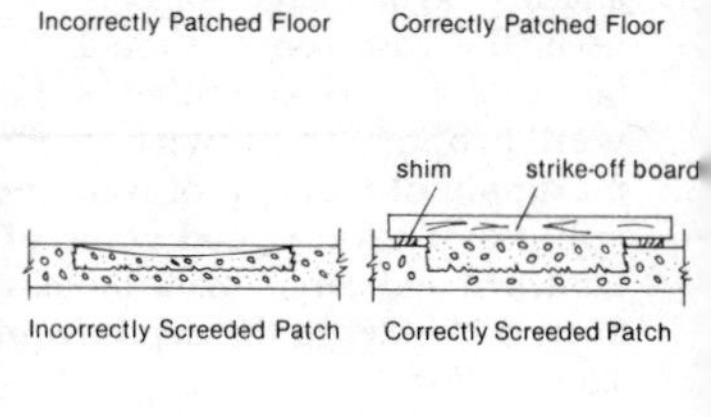

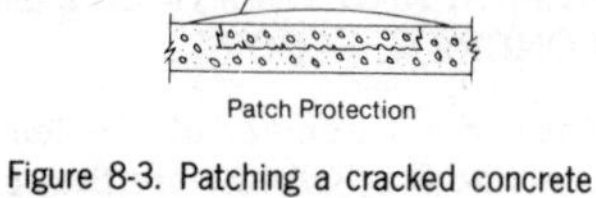

Figure 8-3. Patching a cracked concrete floor.

Roughened Floors. Improperly constructed floors sometimes become pitted or roughened under normal service. In most cases, this can be corrected by grinding off the affected area. However, if the concrete becomes pitted and roughened repeatedly, it is recommended that resurfacing be performed with a good quality concrete.

MASONRY

Properly constructed masonry structures require relatively minimal maintenance and repair needs—cleaning and the repair of any leaking structures.

CLEANING NEW MASONRY

When a new masonry structure has been completed, it is often necessary to remove excess mortar and related stains. The type of cleaning required will be either acid cleaning or acid-free cleaning.

Acid Cleaning. Acid cleaning procedures should be employed where there is excessive staining and imbedded impurities; most newly constructed masonry structures will require it. Before cleaning begins, the maintenance mechanic must be sure that all masonry work is completed, and that the mortar has been thoroughly set and cured. Try to avoid cleaning on hot days, for the acid will not only dry rapidly but will also be absorbed and react quicker with the masonry and mortar. The recommended steps to be followed for acid cleaning are described as follows:

Steps for Acid Cleaning New Masonry

Step 1 *Dry Clean.* All loose and large particles should be removed with wooden paddles and scrapers. Wire brushing and chiseling should be used sparingly to avoid damage to the masonry.

Step 2 *Wall Presoaking.* The wall should then be presoaked with clean water so that it is saturated. At this time, loose particles should also be hosed away.

Step 3 *Mix Acid.* A commercial grade of muriatic acid (hydrochloric acid) should be mixed with water at a 1:9 ratio of acid to water. Only clean fresh acid should be used. Caution: Always add acid to water; never add water to acid.

Step 4 *Apply Acid.* Apply the acid with a long handled fiber brush. Make sure that the area below has been properly cleared. While acid is being applied, it is critical to keep the masonry below soaked to prevent streaking and staining.

Step 5 *Scrubbing.* Once the acid has been applied, scrub the brick, but *not the mortar joints*. Use a stiff fiber brush and wooden paddle for this operation. Do not attempt to clean any area larger than 10 to 20 square feet at one time.

Step 6 *Rinse.* After scrubbing and before the acid has a chance to dry, thoroughly rinse the area with clean water.

Nonacid Cleaning. When the completed structure does not show any heavy soiling or staining, it may be cleaned by using nonacid procedures.

Steps for Nonacid Cleaning New Masonry

Step 1 *Dry Clean.* Remove any large particles of mortar with a wooden paddle and scraper.

Step 2 *Wall Presoaking.* Flush loose material with clean water, and completely saturate the masonry structure.

Step 3 *Scrubbing.* With a stiff fiber brush, scrub down the surface with a solution of 1/2 cup each of trisodium phosphate and a household detergent mixed in one gallon of water.

Step 4 *Rinse.* Once the area has been scrubbed, rinse thoroughly.

CLEANING OLD MASONRY

From time to time, it may be necessary to clean existing masonry. Accumulated dirt from air pollution and spills requires maintenance action. In most cases, this procedure is only required every five to ten years. Listed below are some of the more commonly recommended cleaning methods:

1. *Chemicals with steam* are primarily used to remove stubborn dirt such as paint and other material.
2. *Hand washing* should only be employed for the cleaning of small surface areas.
3. *High cold water pressure* is recommended for cleaning water soluble material.
4. *High pressure steam* is the most popular commercial cleaning method. It is highly recommended for smooth masonry, marble, and other similar materials.

5. *Sandblasting* should only be used where steam pressure cleaning would not succeed, and where there is no possibility of structural damage. Care should be taken when cleaning porous brick, sandstone, and limestone. Since this procedure actually abrades material, a small section should be tested first before sandblasting is approved.
6. *Wet aggregate cleaning* is specifically designed for cleaning limestone, soft brick, and sandstone. The cleaning solution is a nonsilica aggregate mixed with various cleaning solutions.
7. *Wet sand cleaning* should be used on unpolished granite, rock facing, hard bricking materials, or rough finished limestone. Cleaning is accomplished by a mixture of a sand and water spray.

PAINTING

Painting is used to protect and decorate facilities. All paints are made of pigments, or solid powders, and binders, or vehicles. The pigment provides the coloration; the vehicle is the liquid portion of the paint. How often painting needs to be done depends on the specific area to be painted and on atmospheric conditions. When choosing a paint, one should check the manufacturer's recommendations for appropriate use, surface preparation, and methods of application.

EXTERIOR PAINTING

Exterior areas not subjected to abrasion or standing water can be painted with conventional house paints (e.g., water emulsion or solvent-based paints). If industrial fumes are present in the atmosphere, a lead-free paint is recommended to avoid hydrogen sulfide staining.

Paints applied over or next to masonry should be nonchalky to prevent staining. Trim paints requiring color retention should be high gloss, and exterior enamels are recommended for outdoor and metal surfaces over a primer. When painting masonry, a good sealer should be applied prior to painting. For natural exterior coatings, varnishes and oils are recommended. Roofing incorporating galvanized metal requires a special zinc dust-oxide primer under any tough paint.

INTERIOR PAINTING

Interior surfaces should be cleaned and free of any foreign material before paint application. Common types of interior paints and their recommended uses are listed below.

Alkyd paints are weather- and water-resistant and have exceptional gloss and color retention.

Cumar paints are alkali- and acid-resistant and are often used as an aluminum heat-resisting coating.

Epoxy paints provide a very hard finish that is resistant to many chemicals.

Ester gum paints have excellent water- and fade-resistance.

Latex paints are easy to wash and have good color stability.

Maleic paints are fast drying, harder than ester gum paints, and retain their color longer than cumar paints.

Phenolic paints are fast drying. They resist weather, water, and chemicals.

Synthetic rubber paints are excellent for resisting chemicals, alkalis, acids, and abrasion.

PAINTING METAL

One of the most important materials painted in commercial and industrial settings is metal. Table 8-8 provides a guide for applying paint to metal.

TABLE 8–8
Recommended Procedures for the Application of Paint to Metal

Metal Product	Surface Preparation	Recommended Primer	Recommended Paint
Continuous Outdoor Exposure			
Excessive abuse	Remove all loose material, rush, and particles.	Polyamide Epoxy	Polyamide Epoxy
Medium abuse	Remove all loose material, rust, and particles.	Epoxy	Epoxy

TABLE 8–8 (*continued*)

Metal Product	Surface Preparation	Recommended Primer	Recommended Paint
Occasional abuse	Wire brush and remove all loose material.	Keytoxine containing primers	Industrial enamel
Weathering only	Wire brush and remove all loose material.	Keytoxine containing primers	Industrial enamel
Construction equipment	Wire brush or sandblast.	Enamel undercoat	Alkyd enamel
Marine Equipment			
Shore-based products exposed to salt spray	Remove all loose material and rust.	Synthetic rubber	Synthetic rubber base
Parts exposed to fresh water	Wire brush and remove all loose material.	Rubber base	Rubber base
Submerged and exposed boat parts	Wire brush as clean as possible.	Chlorinated rubber	Chlorinated rubber
Indoor Equipment and Furnishings			
Storage facilities and lockers	Wire brush and remove loose materials.	Enamel undercoat	Alkyd enamel
Materials handling equipment	Wire brush and remove loose materials.	Enamel undercoat	Alkyd enamel
Floor trucks, cranes, and powered equipment	Wire brush and remove loose materials.	Enamel undercoat	Alkyd enamel
Machine tools	Wire brush and remove loose materials.	Enamel undercoat	Alkyd enamel

TABLE 8–8 (*continued*)

Metal Product	Surface Preparation	Recommended Primer	Recommended Paint
Equipment components exposed to scuffing	Remove loose material and wash with mineral spirits.	Epoxy	Epoxy
Grillwork, partitions, and canopies	Remove loose material and wash with mineral spirits.	Epoxy	Epoxy
Atmospherically Abusive Conditions			
General acid conditions	Sandblast.	100% solids epoxy	80% solids epoxy
Alkali spray	Sandblast.	100% solids epoxy	80% solids epoxy
Fine abrading particles	Sandblast.	100% solids epoxy	80% solids epoxy
Petrochemicals	Sandblast.	100% solids epoxy	80% solids epoxy
Fabricated Metal Products			
Excessive outdoor exposure	Degrease and chemical etching.	Zinc chromate or chromate-oxide	Alkyd or alkyd-melamine
Flat surfaces	Degrease.	None	Alkyd-urea or alkyd-melamine
Deep drawn or curved surfaces	Degrease.	None	Alkyd-urea or alkyd-melamine
Sanitary products	Degrease and chemical etching.	Epoxy or phenolic rust inhibiting chromate-oxide	Epoxy, vinyl polyurethane, polyester

TABLE 8–8 (*continued*)

Metal Product	Surface Preparation	Recommended	
		Primer	Paint
High temperature exposure	Degrease and chemical etching.	Silicon-alkyd or pure silicon	Silicon-alkyd or pure silicon
Low temperature exposure	Degrease and chemical etching.	Vinyl rust inhibitor	Vinyl
High gloss finish	Degrease and chemical etching.	Chromate-oxide	Alkyd-melamine or alkyd-urea

ELECTRICAL EQUIPMENT MAINTENANCE PROCEDURES

Since many machines and devices make use of electrical and mechanical components, there is, in many cases, an overlapping of electrical, mechanical, and service equipment procedures. This section discusses those procedures that are primarily associated with electrical equipment.

BATTERIES

The most important procedure associated with battery maintenance is charging. If correctly charged, a battery may have a service life in excess of 2,000 charges. Proper charging occurs when the battery has been charged without overheating, overcharging, or generating excessive gases.

LEAD-ACID BATTERIES

There are two major ways to charge lead-acid batteries: the modified constant voltage method and the two-rate method. Other, less common, charging methods include boost, equalizing, and emergency charging.

Modified Constant Voltage Charging. In the modified constant voltage method a constant voltage source is used to supply power to the battery. The batteries are connected in series with a resistor. (See Figure 8-4.)

In the schematic diagram, the constant power source is identified by the dashed line. A generator is used in this example, but a rectifier or DC bus can also be employed if the voltage is stable. Note that the voltage supplied must be equal to 2.63 times the number of cells in the series being charged.

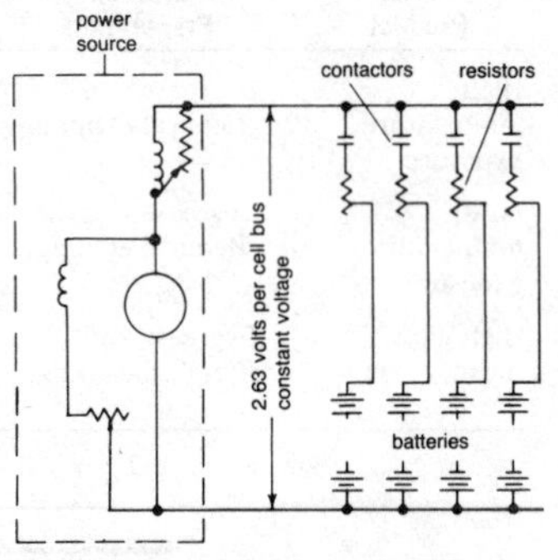

Figure 8-4. Modified constant voltage method of battery recharging.

The batteries are connected in series with resistors used to adjust the rate of charging. The rate should be checked against manufacturer's specifications. Within certain parameters, recharging time can be reduced by decreasing resistance. As the battery becomes fully charged, there will be a tapering off of current flowing to the battery; this indicates that charging occurs at a slower rate near the end of the charging cycle.

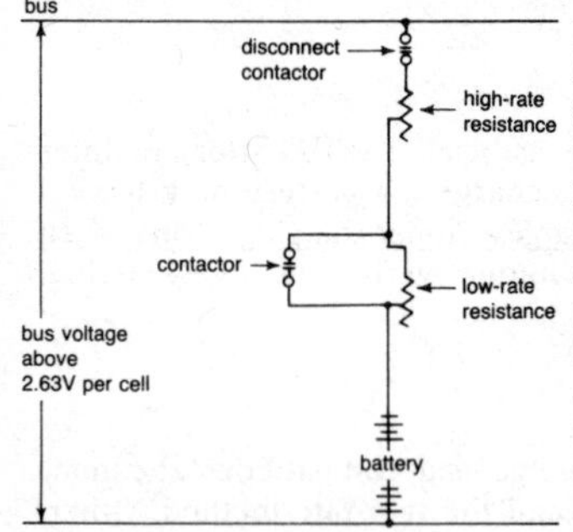

Figure 8-5. Two-rate method of battery recharging.

Two-Rate Charging Method. The two-rate charging method uses a high voltage charge at the beginning and a lower charge at the end of the cycle. The lower charge is considered safer as the battery becomes fully charged. A schematic of this arrangement is provided in Figure 8-5.

In the two-rate method schematic, the voltage must be at least 2.63 times per battery cell and supplied between buses by a power source, such as a generator or rectifier. At first, the low rate resistance is shorted out by the contactor. As the battery becomes charged, the contactor opens to increase the resistance in series with the battery.

NICKEL-CADMIUM ALKALI BATTERIES

The charging of nickel-cadmium alkali batteries is important to maintain a reliable and regular source of power. Unlike lead-acid batteries where charges can be easily and rapidly checked, nickel-cadmium alkali batteries should be checked on a regular schedule.

These batteries can accept overcharging as long as the maximum temperature is held at 115°F; occasionally, these batteries can withstand temperatures as high as 125°F.

Nickel-cadmium alkali batteries are recharged by either constant voltage or constant current charging methods.

Constant Voltage Charging. With the constant voltage charging method, the voltage is usually automatically reduced below recommended levels at the start of the charge. Then, when the battery is capable of handling the charge, the voltage level is increased. Pocket plate cells should be charged at 1.6 volts per cell; power cell and sintered cells at 1.55 volts per cell. Total charging time is approximately 7 hours.

Constant Current Charging. To fully charge a battery with the constant current method, the charger has to be able to deliver a voltage of at least 1.8–1.85 volts per cell. A variable resistance is used in series with the battery to hold the charge steady and make sure that overcharging does not occur.

MOTORS

Electric motors will seldom be damaged as a result of normal operation within rated specifications. The AC motor is less susceptible to damage than the DC motor, because the current-carrying parts are usually more protected in AC motors. However, there are times when some form of damage does occur.

CARE OF THE COMMUTATOR

Of all DC motor parts, the commutator is perhaps most vulnerable to damage. For the commutator to function properly, its surface has to be smooth, concentric, and correctly undercut. One of the most common problems associated with the commutator is nonconcentricity. Machines with peripheral speeds above 9000 rpm, require concentricity to within 0.0005 inch; peripheral speeds between 5000 and 9000 should be concentric to within 0.001 inch. For low speed, large diameter commutators, concentricity should be within 0.003 inch.

When commutators are not concentric, corrective actions should be taken. Grinding is the best method for getting the commutator to within concentric limits.

Grinding Commutators to Concentricity

Step 1 *Grinding Jig.* Most situations require the use of a grinding jig as a mounting method. When possible, grinding should be executed with the armature in its own bearings and within rated speeds.

Step 2 *Grinding.* To assure rigidity during grinding, the brush holder bracket should be properly braced. Select the correct grinding stone for the job. When grinding begins try to prevent metal and wheel dust from entering the windings.

Step 3 *Commutator Rotation.* During grinding, the commutator must rotate. To help in this, a motor drive may be used. If low machining speeds are required, it is possible to turn the commutator on a lathe by taking very fine finishing cuts.

Step 4 *Cleaning.* After grinding, completely clean the commutator, including all slots. Bevel the bar edges to remove burrs and any sharp edges.

Step 5 *Undercutting.* Motors that have mica in their commutator slots may require undercutting. This should be approximately 1/16 inch deep, ±1/64 in.

Step 6 *Set Brush Holder Box.* The brush holder box should then be set at the correct angle and distance from the commutator surface. Usual recommended distances range from 1/6 to 3/16 inch. Check motor manufacturer's specifications.

CARE OF INSULATION

Proper maintenance of motor insulation is required for both AC and DC motors. From time to time, motor insulation must be cleaned and dried. Cleaning should begin when the motor has become dirty. If grease and oil are mixed with the dirt, it may be necessary to wash the motor insulation with an approved solvent. The procedures used to clean and dry motor insulation vary with motor designs and types.

Cleaning Motor Insulation

Step 1 *Wiping.* With a clean dry cloth, wipe off the surfaces to be cleaned so that all loose and dry dirt is removed.

Step 2 *Blowing.* Areas that cannot be reached by hand wiping should be cleaned out with compressed air. Be sure that the compressed air is dry and does not exceed 50 pounds per square inch.

Step 3 *Vacuum.* Remove any remaining loose dirt or particles by vacuuming.

Step 4 *Cleaning.* The cleaning of grease, oil, and other materials can be done by several different procedures, each designed to meet specific requirements and conditions.
Solvent Cleaning. The majority of solvents used for electrical cleaning are toxic, so care must be taken. Any grease- and oil-based dirt should be cleaned with petroleum distillates, chlorinated and petroleum solvents, or coal-tar solvents. Never use carbon tetrachloride or any other chloride solvent.
Water Cleaning. Use water to clean motors clogged with mud and other foreign material. Water applied to insulation should not exceed 25 pounds per square inch. Motors with special high-temperature insulation (e.g., silicone) should only be cleaned with water, since many solvent cleaners will attack the special insulation.
Dry Cleaning. A petroleum absorbent powder can also be used to clean grease and oil. After application, the pow-

der should have sufficient time to absorb the oil and grease. The motor should then be vacuum cleaned.

Step 5 *Drying Insulation.* After the motor has been cleaned, the insulation may require drying. Perhaps the best method is use of an external heat source such as an electric heater or hot air furnace. Within recent years, infrared lamps have become widely accepted for drying insulation.

MECHANICAL EQUIPMENT MAINTENANCE PROCEDURES

Most maintenance procedures for mechanical equipment involve the lubrication and cleaning of component parts. Parts that are broken or damaged usually should be replaced.

BEARINGS

Machinists and maintenance mechanics often need to remove, clean, adjust, and reassemble bearings.

REMOVING BEARINGS

Removing a bearing from a shaft is sometimes difficult. If force is required, apply it to the ring with the tighter fit. A regular gear or bearing puller should be used. (See Figure 8-6.) Steady pressure must be exerted so as not to damage the bearing. At times, it may even be difficult to remove the bearing with a puller. When this occurs, be sure the puller is in position, apply pressure, and pour hot oil into the

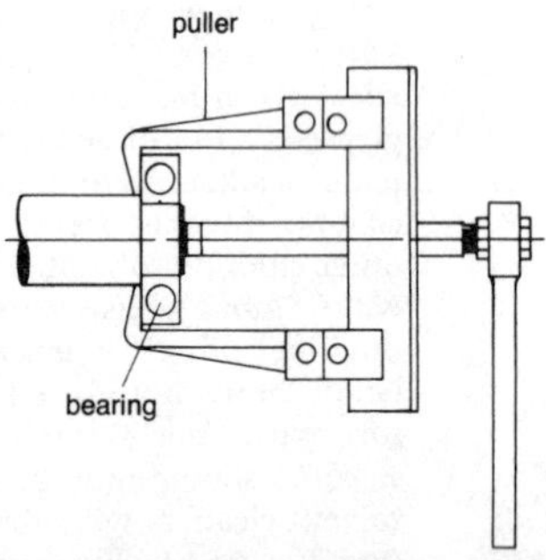

Figure 8-6. Bearing puller.

housing. This will expand the bearing (before the shaft expands) and permit removal. In emergency situations, steam may be used. Dry ice may be brought in direct contact to hollow shafts to accomplish shrinking.

Larger bearings are sometimes removed with hydraulic attachments. As a last resort, it may be necessary to use a hammer and rod to drive the bearing off the shaft. In these situations, the rod should be of a softer material than the bearing.

CLEANING BEARINGS

Once the bearings have been removed from their housing or shafts, they should be cleaned and inspected before they are reinstalled. Cleaning procedures depend on the size of the bearing.

Cleaning Large Bearings. Large bearings such as those used in multiple row arrangements are often washed in cleaning tanks with a hot cleaner. The most common cleaners used are neutral oils (100 sec. @ 100°F viscosity) heated to about 300°F. Alkali cleaners also provide excellent cleansing. Common hot solutions consist of two to three ounces of alkali cleaner (e.g., trisodium phosphate, soda ash, and metasilicate) to one gallon of water.

Cleaning Smaller Bearings. Smaller bearings are usually washed in mineral spirits or kerosene. Other solvents, such as naphtha and gasoline, should not be used because they are volatile. The small cleaning tank should be replenished with solvent frequently.

ADJUSTING BEARINGS

Most bearings cannot be adjusted. However, tapered roller bearings are designed so that they can be adjusted. The advantage of this is that it provides more flexibility in fitting tolerance.

Adjusting Devices. There are a number of devices used to adjust tapered roller bearings. (See Figure 8-7.) *Slotted nut bearing adjusters* are the most common type found in shops. They are used primarily on automobiles and other transporting vehicles. *Double nut and tongued washers* are used primarily on heavy-duty equipment such as trucks and buses. *End-plate and shims* fill the gap be-

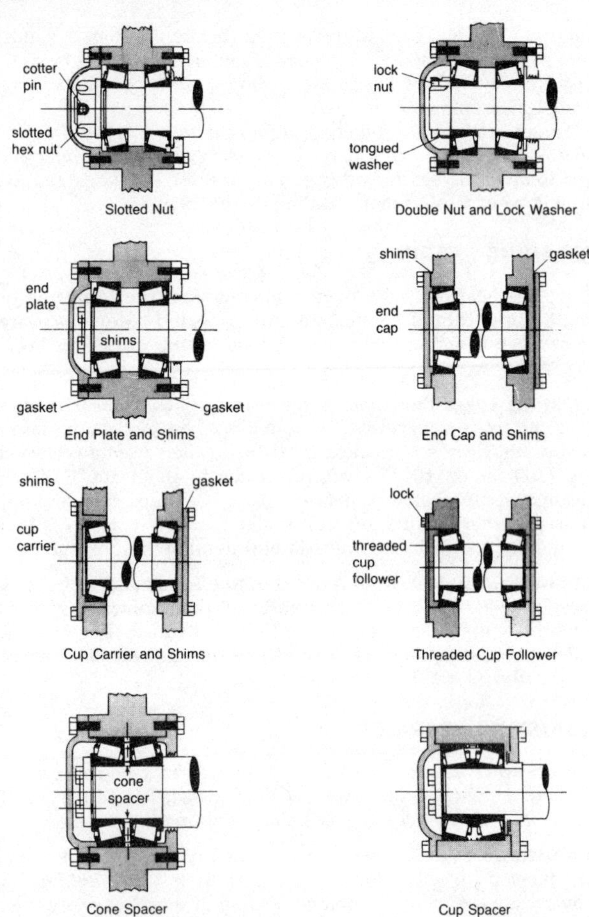

Figure 8-7. Types of bearing adjusting devices.

tween the end plate and the shaft with a shim. *End-cap shims* are used to give the required end play or preload on bearings. *Cup-carrier and shims* use a threaded cup follower which is pulled until the bearings bind and then is backed off. *Cone spacers* allow for adjustments by means of a spacer located between the cones.

ASSEMBLING BEARINGS

When assembling bearings it may be possible to slip them into position by hand. For tighter fittings, a hand press may be used. If the bearing is too large for a hand press, or if it requires additional force, tubing can be placed against the bearing and then struck by a hammer. Figure 8-8 illustrates how force is applied against the inner ring of a bearing by tubing material.

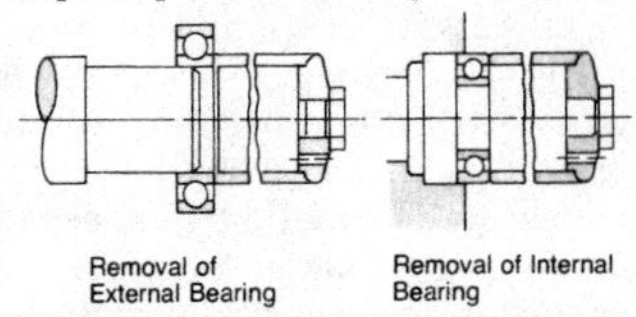

Figure 8-8. Use of tubing for bearing removal.

BELTS AND CHAIN DRIVES

Efficient belt and chain drive performance depends on proper installation, alignment, and tensioning. A procedure that can be followed for aligning shafting and pulleys is illustrated in Figure 8-9.

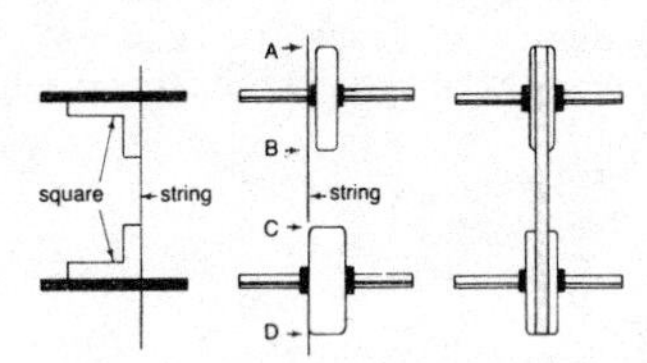

Figure 8-9. Method of aligning shafting and pulleys.

Aligning Shafting and Pulleys

Step 1 With a level, check the shafts for straightness and proper positioning.

Step 2 Using a taut string and squares between the shafts, make sure that the shafts are parallel to each other.

Step 3 Placing a taut string along the edges of the pulleys, determine if they are properly aligned. If the pulleys are the same size, they should touch lightly at four points (see points A, B, C, and D in the illustration). If the pulleys are of different sizes, the distances from the string should be the same along each individual pulley (points A and B).

PACKINGS AND SEALS

Seals, gaskets, seats, and packing are used to prevent leakage at joints and between spaces. Though leakages can occur as a result of defects in housing units, of interest here is the prevention of leaks at static and dynamic joints. A *static joint* is one that is relatively free of motion—for example, fitted joints of continuous piping and tubing or an engine head crankcase cover. A *dynamic joint* is one in which one or more members is in motion relative to other members, as in rotational and/or reciprocating motion.

GASKETS

Gaskets are usually designed to be installed between two stationary and rigid components to prevent fluid flow. They are used in containers designed with parallel flanges or concentric cylinders. When leakage does occur, it can be stopped by tightening the opposing surfaces. (See Figure 8-10.)

Gaskets are available in sheet, strip, or bulk form. Those designed with a plain groove should never protrude beyond the height of the groove (to prevent mushrooming during tightening).

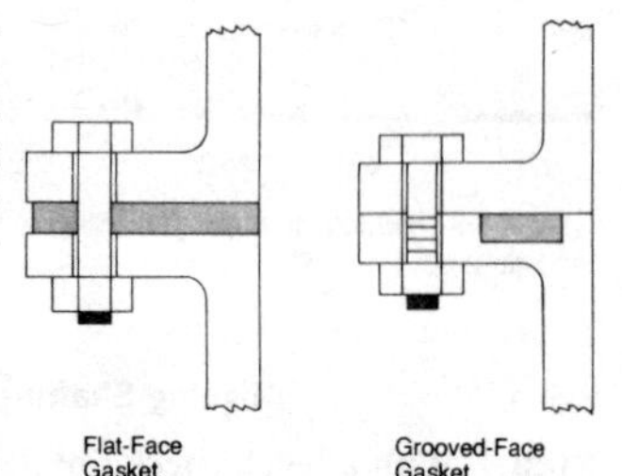

Figure 8-10. Different types of gaskets.

PACKINGS

Packing includes all hand-formed or cut material that fits around joints to prevent leakage. Common types of packing are preformed square, rectangular, conical, cupped, O-, V-, and U-shaped. (See Figure 8-11.)

Packings are used in a variety of machines. *Dynamic packing* is an adjustable device that is installed in a stuffing box. It is adjusted by a gland to the desired fit. In some cases, this type of packing is referred to as a *jamb packing*. A *dynamic seal* is automatic, or nonadjustable. It is installed in a predetermined and specified space and condition.

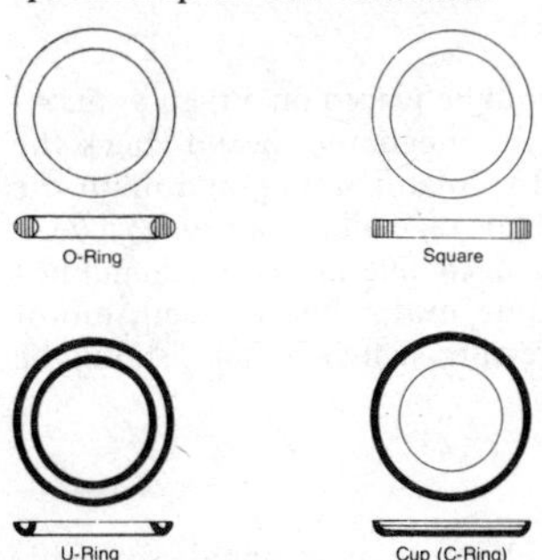

Figure 8-11. Some types of packings.

Regardless of the type used, packings will usually function only under fluid pressure. They are designed to work under close conformity and contact with moving parts. Hence, they function like a bearing. With dynamic packing, no actual contact is established between the surfaces because of the lubricating film required for operation. Without the lubricating film, heat will be produced and damage caused. The film must be of appropriate thinness so that it will not flow through the clearance and cause excessive leakage.

SEALS

Generally, seals differ from packings in that they are used to keep dirt and water out of bearings and lubricants. Rotary shaft seals are used to prevent both the loss of fluids from the system and the entrance of air into refrigerant compressors.

SERVICE EQUIPMENT MAINTENANCE PROCEDURES

Maintenance procedures used for service equipment often represent a substantial part of a maintenance and machine mechanic's job. This section discusses those procedures frequently required of the mechanic.

AIR CONDITIONING SYSTEMS

Electrical equipment found in air conditioning systems has already been addressed. Presented here are step-by-step procedures for the proper care of cooling and heating units.

NEED FOR PERIODIC CHECKS

The system's air compressors should be placed on a regular maintenance schedule. Every week, the mechanic should check the level of oil in the crankcase and regulator setting and drain the tank water and the combination filter-regulator. Four times a year, the crankcase oil should be changed, cooling fins cleaned, and belt tension adjusted. During this time motor bearings oil, motor brushes and commutator, and pressure switch contacts should be checked.

HEAT PUMP

An air conditioning unit that has become more popular over the years is the heat pump. Heat pumps are reversible heating and cooling systems. They are designed with a special four-way valve, and the normal cycle can be reversed with the condenser functioning as an evaporator and vice versa. General maintenance procedures outlined in Chapter 7 apply to heat pumps.

Maintenance Procedures for Heat Pumps

Step 1 Conduct a regular check of the outdoor coil. Because frost can form rapidly, it is important to look for fast and efficient defrosting. If poor defrosting occurs, check the defrost thermostat, reversing valve, defrost relay, wind effect, and the timer.

Step 2 Inspect the check valve for any leakage. If the valve is stuck in a closed position, repair it. Any valve replacement should be executed after pump down.

Step 3 Check the reversing valve. If defective, do not repair: it should be replaced.

Step 4 Replace air filters at least four times a year.

Step 5 Clean indoor and outdoor coil surfaces.

Step 6 Check refrigeration piping for leaks. Make sure that there is no vibration against any surface.

Step 7 Clean and inspect the indoor coil condensate drain and the unit's mounting.

Step 8 Check all wiring and contacts. Check and adjust fan belt tension.

Step 9 Check defrost thermostat and clean the contacts. Also check crankcase heater terminals.

CARE OF COMPRESSOR AND REFRIGERANT

The compressor, the motor driving the compressor, and the refrigerant all require careful maintenance. During the colder months, outside air should be used for cooling. When the temperatures drop in the fall, the refrigeration system should be pumped down and the refrigerant valve turned off in the condenser. This will confine the refrigerant to an area with a small number of joints and minimize refrigerant loss.

When the cooling system is no longer needed, belt driven compressors require that the motor be moved toward the compressor to lessen belt tension. This prevents the belts from setting flat spots. When spring arrives, the motor should be repositioned.

Maintenance Procedures for Towers and Evaporative Condensers

Step 1 Check the condensing coil and remove any scaling.

Step 2 Check and clean spray nozzles.

Step 3 Flush and clean pump strainer, water strainer, and air intake screen.

Step 4 Inspect and make necessary repairs to float controls.

SEASONAL LAY-UP PROCEDURES

Typically, seasonal lay-up procedures take the units out of service for one to seven days. A one week maximum is usually adhered to because within that time span serious corrosion is unlikely.

Lay-Up Procedures for Forced Air Circulation Ducts and Registers

Step 1 Check and repair any deteriorated areas of insulation around the ductwork.

Step 2 Clean all registers and diffusers.

Step 3 Remove grills and vacuum by lowering vacuum cleaning nozzle into the duct to collect accumulated dirt.

Step 4 Wipe clean all wall and ceiling smudges.

Lay-Up Procedures for Steam Boilers and Hydronic Systems

Step 1 Drain boiler and wash out thoroughly with water under pressure.

Step 2 Clean gas passages, breeching, and damper with scraper and wire brush.

Step 3 Inspect and repair all outside surfaces.

Step 4 With oily rags, swab the tubes of fire-tube boilers from one end to the other.

Step 5 Remove and clean the gage glass.

Step 6 Place a wooden tray of slaked lime inside water space and close up all manholes and openings.

Lay-Up Procedures for Direct Fired Space Heaters

Step 1 Clean and adjust firing equipment.

Step 2 Clean all passages in the space heater and vacuum out all dust and lint. If system is floor furnace, be sure to cover all gratings to prevent dust entry during shutdown.

Step 3 For oil fired units, check oil supply valve, making sure it is closed tight at the tank.
In gas fired units, if bottled gas is used, check the gas valve to make sure it is closed at the tank. If the unit is fed directly by a gas line, have the utility company shut off the gas supply.

Lay-Up Procedures for Direct Fired Unit Heaters

Step 1 Clean soot from heat exchange areas.

Step 2 Blow out dust with compressed air hose.

Step 3 Clean and adjust firing equipment.

Step 4 Inspect fan blades for balance affecting deposits and remove if found.

Step 5 Lubricate motor and fan bearings.

Lay-Up Procedures for Gas Fired Infrared Heaters

Step 1 Check, clean, and make necessary adjustments in the gas burning system.

Step 2 Enclose the unit with covering to prevent dust from entering and clogging the orifices.

Lay-Up Procedures for Steam and Hot Water Units

Step 1 Drain unit and let stand until dry.

Step 2 Clean heat exchange surface by blowing with compressed air.

Step 3 Encase heater with appropriate covering in dirty and/or corrosive atmospheres.

Step 4 Check and clean fan blades of unbalancing deposits.

Step 5 In steam units, clean trap.

FLUID POWER SYSTEMS

Maintaining fluid power systems involves three simple procedures that will greatly affect performance, efficiency, and service life.

1. Be sure that hydraulic fluid of the proper type and viscosity is kept clean and in sufficient quantity.
2. Change filters and cleaning strainers periodically.
3. Be sure that all connections are kept tight (but not excessively tight).

The basic method recommended for changing oil in fluid power systems follows:

Step 1 Drain the system completely.

Step 2 Remove the reservoir clean-out cover and swab and clean out the bottom of the reservoir.

Step 3 Remove the suction filter and soak it in solvent (kerosene) overnight. Change the filter element. Before reinstallation, blow out the filter with compressed air to remove any remaining solvent.

Step 4 In systems that have a significant amount of dirt, introduce sufficient amounts of oil into the reservoir to operate the system slightly above minimum fluid level. Circulate for 15 to 20 minutes through the system; then drain reservoir again. This process is known as a *short fill*.

PIPING SYSTEMS

Maintenance and repair procedures for piping systems are applicable to water, gas, and steam systems; fluid power systems; air conditioning systems; and industrial machines and equipment. Pipe and pipe products are available in several common materials, such as polyvinylchloride (PVC), copper, cast iron, and brass. Most maintenance considerations are similar, regardless of the material used.

One of the most important duties a maintenance mechanic may face is that of emergency repairs. There are times when a pipe leakage must be repaired on the spot—without time for shutting down the entire system. (See Figure 8-12a.)

Emergency Repair for PVC and Cast Iron Pipe

1. Within the area of the crack, apply cement (for example, vinyl for PVC piping and iron cement for cast iron pipes), and bind it tightly with sheet metal.
2. From a next-size-larger pipe, cut a half shell and seal with cement or a soft gasket.
3. When a pipe joint is needed, slip the ends of the pipes into a larger pipe and caulk with cement.

The recommended repair procedure for copper and brass piping is different from that for PVC and cast iron.

PIPE THREADS

Some pipe products are designed with threads. The Standard pipe thread is used for most situations; Extra Strong and Double Extra Strong are used where greater stresses exist. These threads may be prefabricated by the manufacturer or cut in the maintenance shop. There are times, however, when the threaded section of pipe will become damaged or broken. (See Figure 8-12b.)

PIPE VALVES

Within the piping system of a plant, there are numerous sets of valves used to control the flow of water and other fluids. One of the most common valves encountered is the *globe valve*. From time to time, it may be necessary to regrind the valve seats. The simplest procedure for grinding globe valve seats is to pin the stem and disk together so that the bonnet is used as a guide for lapping or grinding. (See Figure 8-13.)

Grinding Globe Valve Seats

Step 1 Lock the disk and stem together by rotating the disk so that it lines up the locknut slot with the hole in the stem. Locking is accomplished by inserting a pin or small finishing nail.

Step 2 With your finger, apply a thin coating of grinding compound to the seat and disk. The compound used should be fine-grained.

Step 3 Place the bonnet assembly into the valve body and rotate by hand until the disk just touches the seat. Continue to turn the unit until the bonnet rises about 1/32 inch.

Step 4 Screw on the unit handtight, and then back off one complete revolution. The hand wheel and bonnet should be oscillated, and a new spot occasionally turned to.

Step 5 Once grinding is completed, wipe the seat and disk completely with a soft cloth wetted with a solvent (e.g., gasoline or kerosene).

Step 6 Remove the locking pin from the hole and reassemble the valve.

Emergency Repair for Copper and Brass Pipe

Step 1 Drain the pipe and stop the flow.

Step 2 Cut an appropriately sized copper patch and shape it to fit the damaged pipe.

Step 3 Clean the mating surfaces with an abrasive cloth and hydrochloric acid.

Step 4 Bind the patch in place securely with wire.

Step 5 With bricks, enclose the damaged area to confine the heat, and heat with an acetylene torch. Do not burn or overheat the patch or pipe.

Step 6 Run solder with borax flux between the surfaces. Work the area so that the solder will flow between all parts.

Step 7 Test the patch by allowing the pipe contents to flow under pressure.

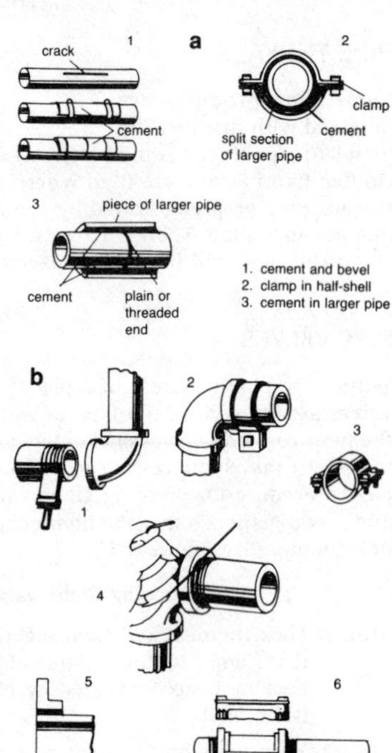

Figure 8-12. Techniques for emergency repair of pipes and joints.

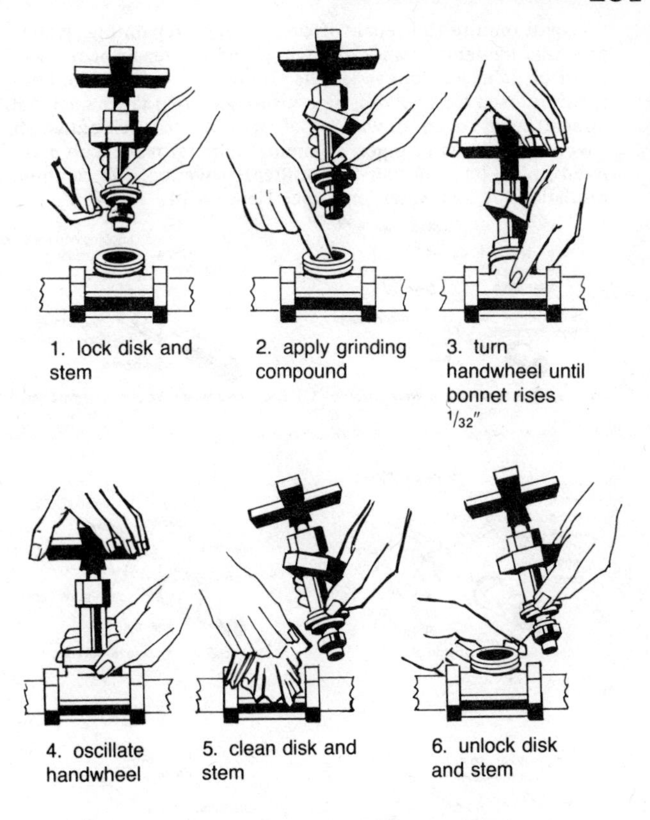

Figure 8-13. Steps in regrinding globe valve seats.

PIPE INSULATION

Routine maintenance of piping systems should include inspection of warm and hot pipe insulation. From time to time, the insulation will require the repair of cuts or tears, repainting, waterproofing, tightening bands and wires, and the repair or replacement of damaged canvas jackets. There are six basic methods typically used for applying pipe insulation: two layer sectional mineral wool covering, waterproof jacketing, hair-felt antisweat covers of cold water pipes, segmental mineral wool with a cement finish, built up hair-felt antifreeze covering, and common insulation for cold water lines. (See Figure 8-14.)

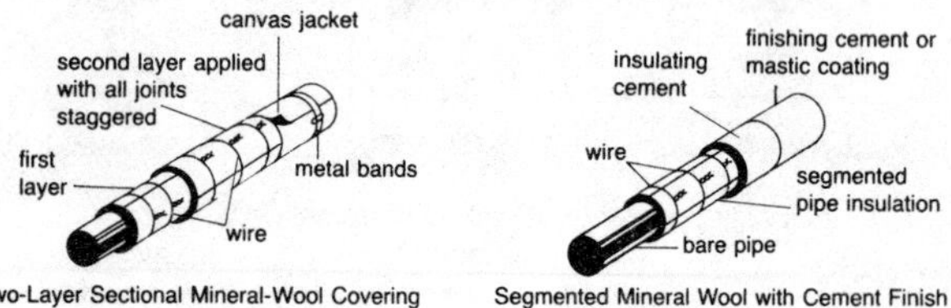

Two-Layer Sectional Mineral-Wool Covering

Segmented Mineral Wool with Cement Finish

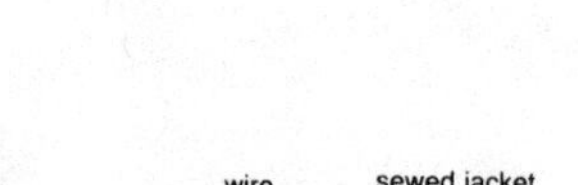

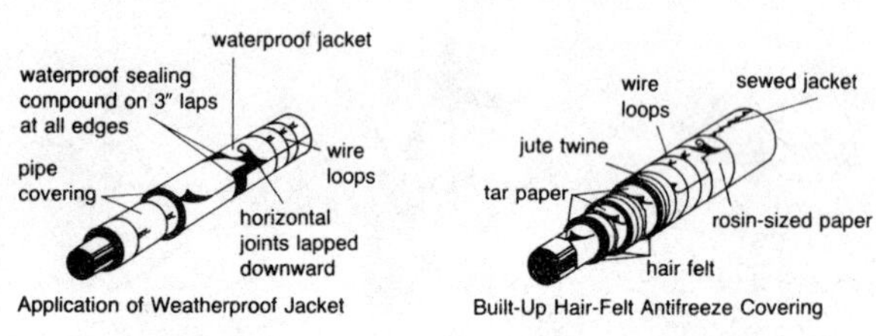

Application of Weatherproof Jacket

Built-Up Hair-Felt Antifreeze Covering

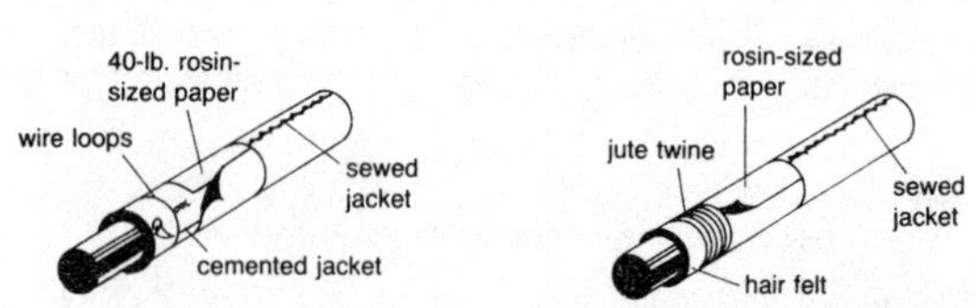

Hair-Felt Antisweat Cover for Cold Water

Typical Insulation for Cold Water Lines

Figure 8-14. Techniques for installing pipe insulation.

PART FOUR

MAINTENANCE FUNDAMENTALS

9

Basic Mathematics for Mechanics

In addition to technical knowledge and skills, the mechanic must also be proficient in a number of related disciplines, one of which is basic mathematics. Math is used every day to help solve problems and make decisions. It is used to select the correct size components and devices, to determine machine capacities, and to make correct equipment settings. This chapter briefly reviews the basic math needed by machine and maintenance mechanics for accurate and precise work.

NUMERALS AND SYMBOLS

The system now used throughout the world for expressing numerical values is the Arabic system. This system makes use of ten numerical symbols: 0, 1, 2, 3, 4, 5, 6, 7, 8, and 9. From these, any value can be expressed, including whole numbers, fractions, and decimals.

WHOLE NUMBERS

Whole numbers are the basis of all mathematical procedures. In many cases, they are not used alone, but in combination with fractions and decimals. Critical here is the placement of each number so that it will represent a desired quantity. Thus, the term "place" has an important meaning in mathematics. The meaning of the place of numbers can be easily explained in the number: 874,365,432.

This number is read: "eight hundred seventy-four million, three hundred sixty-five thousand, four hundred thirty-two." Starting from the right, the first three digits (432) represents units, followed by thousands (365), and then millions (874). The number furthest right in each group symbolizes a single denomination of that place (e.g., two, five thousand, and four million). The middle digit represents a ten multiple (e.g., 30, 60 thousand, and 70 million). The third digit represents multiples of a hundred (e.g., 400, 300 thousand, and 800 million).

FRACTIONS

In many mathematical calculations whole numbers are used in combination with a value that is part of a whole number. One of the most common methods used for expressing these values is the fraction. Fractions are typically used when great accuracy is not required—for example, in carpentry, construction, and piping work, where measurements are seldom smaller than 1/64 inch.

All fractions are made up of two numbers: a numerator and denominator. The numerator is the top number of the fraction; the denominator is the bottom number. In the fraction 7/16, 7 is the numerator and 16 is the denominator.

DECIMALS

Maintenance and repair work on machines and equipment often requires more precise measures than fractions can offer. A number expression easier to use than fractions is the decimal. Decimals are used in most gages, meters, micrometers, and other measuring devices.

A decimal is a division of one. It is preceded by a *decimal point.* To the left of the decimal point are whole numbers, and to the right of the decimal point is the division of one. Similar to whole numbers, the relative position of a number to the right of the decimal point determines its exact value. These values are based upon the base number ten. For each position away from the decimal point, one multiplies by ten. For example, the numerical values for each number in the decimal 0.2549 are:

$$2 = 2/10$$

$$5 = 5/100$$

$$4 = 4/1000$$

$$9 = 9/10{,}0000$$

$$0.2549 = 2549/10{,}000$$

NOTATIONS AND SYMBOLS

There are several common formula notations and symbols found within the trade. These formulas are used to obtain specific values that are useful in calculating quantities such as volume, area, weight, and cost per unit. In most cases, these notations and symbols are employed to represent a given mathematical procedure.

EXPONENTS (POWERS)

One of the most common notations is the *exponent,* or *power,* note, which is expressed as a superscript. Superscripts are numbers or letters (i.e., alpha-numerics) that are placed on the upper right side of another alpha-number such as 4^2, x^3, and 4^b. Exponents are used to show how many times a number is to be multiplied by itself. Thus:

$$7^1 = 7 \times 1$$

$$7^2 = 7 \times 7$$

$$7^3 = 7 \times 7 \times 7$$

$$7^4 = 7 \times 7 \times 7 \times 7$$

In exponential expressions, a number with an exponent of two is said to be squared or raised to the second power; a number with an exponet of three is cubed or raised to the third power. After the exponent three, the power expression is used—fourth power, fifth power, sixth power, etc.

Some formulas make use of negative exponents. The number raised to a negative power is equal to the reciprocal of that number with a positive sign. Examples of this are:

$$4^{-2} = 1/4^2 = 1/16$$

$$y^{-4} = 1/y^4$$

In some cases, one will be required to use formulas with a zero exponent. Any number, except for zero, raised to the power of zero is equal to 1. Thus $4^0 = 1$ and $98^0 = 1$.

SCIENTIFIC NOTATION

Notation by powers of ten is often used to indicate the exact position of the decimal point. This is known as *scientific notation.* This notation provides for both positive and negative powers of ten. When the power of ten is positive, the decimal point of the number is moved to the right as many spaces as the exponent. Thus 3.1416×10^5 equals 314,160, because the decimal point is moved to the right five spaces. When the power of ten is negative, the decimal point is moved to the left. Thus the expression 3.1416×10^{-8} means 0.000000031416.

ROOT EXPRESSIONS

Another notation commonly used within the trade is the root expression, which is represented by the radical sign: $\sqrt{\ }$. The index of the root is located at the upper left corner, such as the three in $\sqrt[3]{\ }$. If the index has no number, then it is assumed to be 2, representing the *square root* of the number under the radical sign. The square root of a number is that number which multiplied by itself equals the original value. For example, $\sqrt{16}$ (the square root of 16) is equal to 4. $4 \times 4 = 16$. $\sqrt[3]{64}$ is the cube root of 64 which equals 4. $4 \times 4 \times 4 = 64$.

GREEK SYMBOLS

An important notation used in a number of different formulas is the Greek letter pi, represented by the symbol π. Unlike many symbols, pi has a specific numerical value which is 22/7. Because 22/7 can never be expressed evenly as a decimal (22 cannot be divided evenly by 7), some favor using the fraction value in their calculations. A more common practice is to use 3.14 for the value of pi. More precise calculations use 3.14159 as the value of pi.

ALGEBRA

Algebra is used as an extension of arithmetic for the solving of many problems. In addition to numbers, algebraic expressions also make use of letters to express value. The vast majority of formulas used in the mechanic trade use algebraic expressions.

Both upper case and lower case letters are used for algebraic symbols. In addition to the English alphabet, Greek letters are used in many formulas. When Greek letters are encountered, one should not become confused, for that letter functions as other letters—to represent a given value. For example, the Greek letter pi (π) is the constant 22/7 or 3.14159.

Mathematical procedures are the same for algebraic expressions as for arithmetic. For example, if the two nonnumerical variables x and y are added together, subtracted, multiplied, or divided, they would be represented in the same manner as if they were numbers:

addition: $x + y$

subtraction: $x - y$

multiplication: xy or $(x)\ (y)$

division: x/y

In algebra, parentheses ( ), brackets [ ], and braces { } are used to indicate the order in which calculations are to be performed. In the expression:

$$\frac{(x - y)}{(a - d)} = f$$

we must first subtract y from x and then multiply that difference by z. Next, d must be subtracted from a before that difference can be divided into the numerator. Once divided, the value of f will be determined. (In other words, calculations within parentheses should be performed first.)

Exponents used in algebraic expressions follow the same rules as used in working numbers. Whole positive exponents indicate a multiplying factor; negative powers are reciprocal quantities. In some formulas, fractional exponents are used to indicate the root of a value. Examples of this are:

$$x^{1/2} = \sqrt{x}$$
$$x^{1/3} = \sqrt[3]{x}$$
$$x^{1/n} = \sqrt[n]{x}$$

TRIGONOMETRY

Trigonometry is used when working with triangles, which may be either right (having one 90° angle) or oblique (no 90° angles). (In all triangles, the sum of the three angles is 180°.) Trigonometric calculations are often adapted to slanting surfaces and the positioning of machine elements and cutting tools, as well as to the location of fasteners and weldments. The formulas used for trigonometric problems are based on algebraic expressions.

RIGHT TRIANGLES

All right triangles are made up of two acute angles (less than 90°) and one right angle (90°), where the sum of all three angles equals 180°. When solving problems involving right triangles, refer to the Table of Natural Trigonometric Functions, found in most mathematics handbooks and references. Today, most calculations are made with the aid of hand calculators that are programmed with these functions, so that the trigonometric functions table is not required. (See Figure 9-1.)

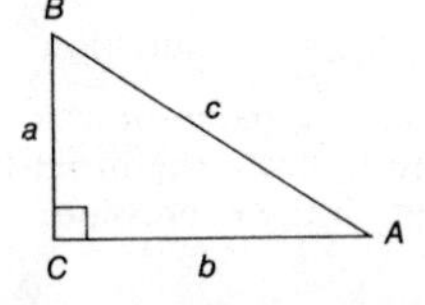

Figure 9-1. Right triangle ABC.

When working with trigonometric formulas, one will encounter functions that are numerical values expressed as sine (sin), cosine (cos), tangent (tan), and cotangent (cot). The functions for different-sized angles are found in the Table of Natural Trigonometric Functions, and can be calculated by the following formulas:

$$\tan A = a/b = \text{opposite/adjacent}$$
$$\sin A = a/c = \text{opposite/hypotenuse}$$
$$\cos A = b/c = \text{adjacent/hypotenuse}$$

DETERMINING ANGLE SIZE

If two sides of a right triangle are known, any of its angles can be calculated by use of these formulas. For example, assume that we have a right triangle whose sides are $a = 30''$, $b = 40''$, and $c = 50''$. To find the size of angle A, we use the following procedures:

$$\tan A = a/b = 30/40 = .7500$$

Using the Table of Natural Trigonometric Functions, or an appropriate calculator, we find that the angle whose tangent is equal to .7500 is about 36° 52′. Other trigonometric functions can also be used to calculate the same angle:

$$\sin A = a/c = 30/50 = .60000 = 36° 52'$$
$$\cos A = b/c = 40/50 = .800000 = 36° 52'$$

DETERMINING LENGTH MEASURES

Formulas are also used to calculate unknown length measures. The formulas used when two sides of a right triangle are known are as follows:

$$c^2 = a^2 + b^2$$
$$a = \sqrt{(c - a)\ (c + b)}$$
$$b = \sqrt{(c - a)\ (c + a)}$$
$$c = \sqrt{a^2 + b^2}$$

Other right triangle formulas that are used when one angle and one side are known are listed here:

$$a = (\sin A)(c) = (\cos B)(c) = (\tan A)(b)$$
$$= (\cot B)(b) = b/\cot A = b/\tan B$$
$$b = (\cos A)(c) = (\sin B)(c) = (\cot A)(a)$$
$$= (\tan B)(a) = b/\tan A = b/\cot B$$
$$c = a/\sin A = a/\cos B = b/\cos A = b/\sin B$$

OBLIQUE TRIANGLES

An oblique triangle is a triangle that has no right angle. Similar to right triangles, however, missing dimensions can be calculated by using trigonometric formulas. These formulas are based upon the Laws of Cosines, Sines, and Tangents. The formulas derived from these laws are described below. (See Figure 9-2.)

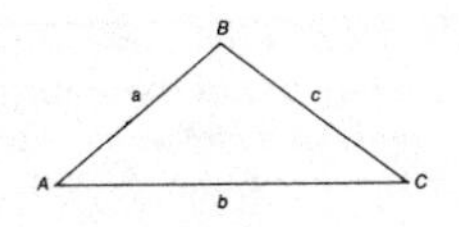

Figure 9-2. Oblique triangle ABC.

TRIGONOMETRIC PROBLEMS

One of the more important calculations used in the trade is the area (A) of a triangle. This problem is used in a variety of situations, including material cost calculations. The basic requirement is that two sides and an included angle be known. The three formulas that are used for determining the area of a triangle are:

$$A = .5bc \sin A$$
$$A = .5ab \sin C$$
$$A = .5ac \sin B$$

A list of trigonometric problems commonly encountered in the trade and recommended solutions are as follows:

1. If you know two angles and one side and want to determine the lengths of the other sides, use the Law of Sines.
2. If you know two sides and an opposite angle, use the Law of Sines to determine the other side and angles (though it is an ambiguous problem).

3. If you know two sides and the included angle and want to determine the third side of the triangle, use the Law of Tangents to find the other angles, followed by the Law of Sines to determine the length of the third side of the triangle.
4. If you are given the three sides and need to determine the sizes of the angles, use the Law of Cosines.

BASIC TRADE FORMULAS

Problems involving the calculation of surface area and solid volume are frequently encountered by machine and maintenance mechanics. Area and volume problems are important in determining such factors as storage capacity, shipping weights, and material costs. All formulas used here are based on the principles of geometry.

AREA FORMULAS FOR PLANE FIGURES

The ability to solve simple area problems of plane figures is central to more complex problem solving. The formulas presented here are used in the trade when dealing with plane, or flat, surfaces.

SQUARE

A square is a parallelogram with four equal sides that are perpendicular to each other. (See Figure 9-3.) To calculate the area (A) of a square, two formulas are available. Which formula to use depends on the known dimensions. The first formula is used when the sides (s) of the square are known; the second is used when the square's diagonal dimension (d) is known. These formulas are:

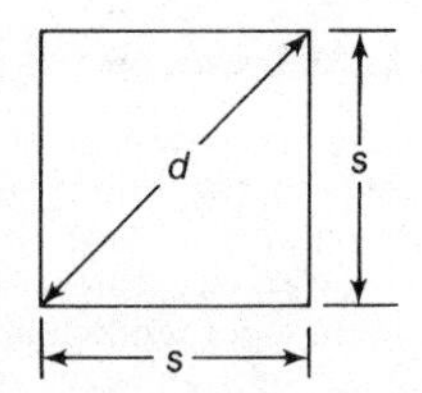

Figure 9-3. Square.

$$A = s^2$$
$$A = d^2$$

When the area of the square is known, the length of a side can be determined by taking the square root of the area ($s = \sqrt{A}$).

When the diagonal is known, the length of a side can be determined by the formula $s = 0.7071d$.

RECTANGLE

A rectangle is a parallelogram whose sides are perpendicular to each other, and in which the two pairs of sides are of differing dimensions. (See Figure 9-4.) Similar to a square, the area formulas for a rectangle require either side or diagonal dimensions.

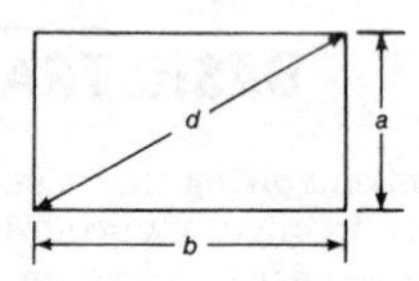

Figure 9-4. Rectangle.

$$A = ab$$
$$A = a\sqrt{d^2 - a^2}$$
$$A = b\sqrt{d^2 - b^2}$$

Other formulas that incorporate dimensional and area knowns can be used to determine the unknown dimension.

$$a = A/b = \sqrt{d^2 - b^2}$$
$$b = A/a = \sqrt{d^2 - a^2}$$
$$d = \sqrt{a^2 + b^2}$$

PARALLELOGRAM

A parallelogram is a plane figure in which the opposing sides are parallel to each another. In our example the sides are not perpendicular. (See Figure 9-5.) The area of a parallelogram is calculated by multiplying the dimension of the base side by its height.

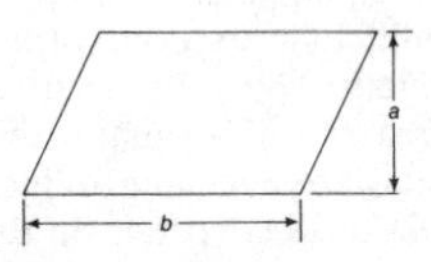

Figure 9-5. Parallelogram.

$$A = ab$$
$$a = A/b$$
$$b = A/a$$

TRIANGLES

A triangle is any three-sided plane figure. There are three categories of triangles: right, acute, and obtuse. Right triangles have one angle that is 90°; an acute triangle is one where all angles are less than 90°; an obtuse triangle is one that has one angle greater than 90°. (See Figure 9-6.)

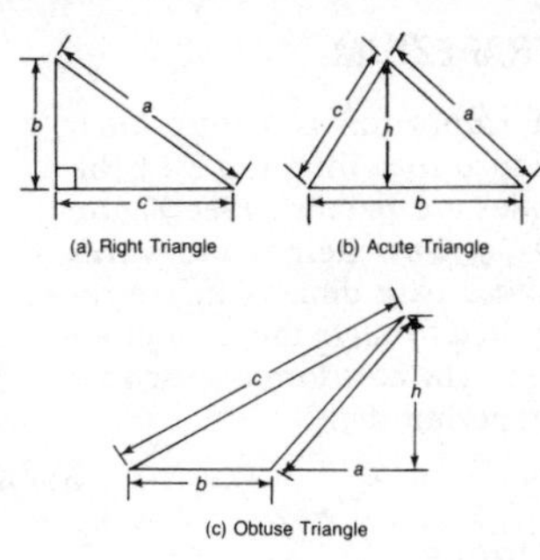

Figure 9-6. Categories of triangles.

Area formulas used for each type of triangle follow:

Right Triangle Formulas

$$A = bc/2$$
$$a = \sqrt{b^2 + c^2}$$
$$b = \sqrt{a^2 - c^2}$$
$$c = \sqrt{a^2 - b^2}$$

Acute Triangle Formulas

$$A = bh/2$$
$$A = (b/2)\sqrt{a^2 - [(a^2 + b^2 + c^2)/2b]^2}$$

Obtuse Triangle Formulas

$$A = bh/2$$
$$A = (b/2)\sqrt{a^2 - [(c^2 - a^2 - b^2)/2b]^2}$$

TRAPEZOID

A trapezoid is a four-sided figure that has only two parallel sides (see Figure 9-7). Area calculations make use of parallel side and height dimensions. The formula recommended for determining the area of a trapezoid is:

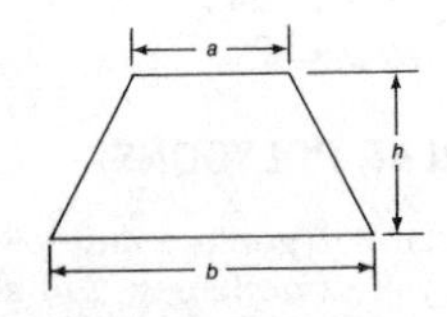

Figure 9-7. Trapezoid.

$$A = (a + b)h/2$$

TRAPEZIUM

A trapezium is a four-sided figure in which none of the sides are parallel. (See Figure 9-8.) Two height measures and a base dimension are required to solve this area problem. The area formula used for trapezium figures is:

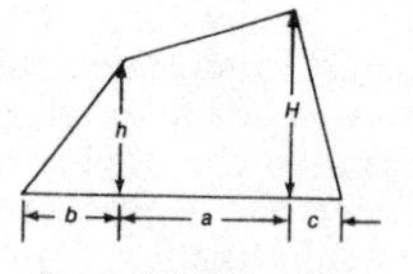

Figure 9-8. Trapezium.

$$A = [(H + h)a + bh + cH]/2$$

CIRCLE

Calculations made for circle specifications are important in many jobs (see Figure 9-9.) There are two important circle formulas: circumference (C) and area (A). These and related formulas follow:

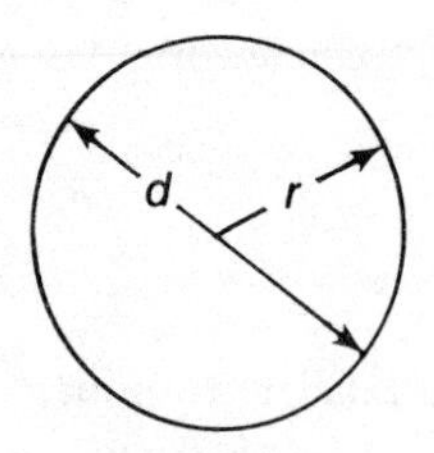

Figure 9-9. Circle.

Circumference Formulas

$$C = 2\pi r$$

$$C = \pi d$$

$$r = C/6.2832$$

$$d = C\pi$$

Area Formulas

$$A = \pi r^2$$

$$r = 0.564\sqrt{A}$$

$$r = \sqrt{A/\pi}$$

REGULAR POLYGONS

A regular polygon is a multiple-sided figure in which all the sides are of the same length and all angles are equal. Two common regular polygons are hexagons (six equal sides) and octagons (eight equal sides). (See Figure 9-10.) Area and related formulas for each are listed on the next page.

Regular Hexagon Formulas

$$A = 2.598s^2$$
$$A = 2.598R^2$$
$$A = 3.464r^2$$
$$R = s$$
$$s = 1.155r$$
$$r = 0.866s$$
$$r = 0.866R$$

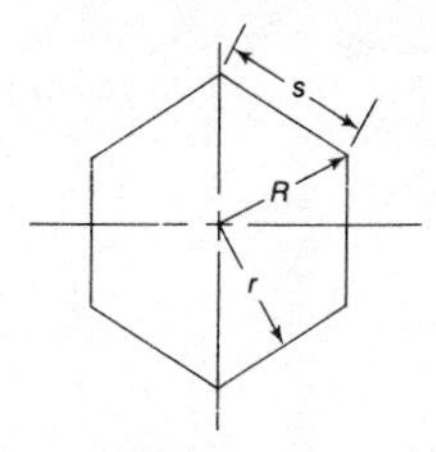

(a) Regular Hexagon

Regular Octagon Formulas

$$A = 4.828s^2$$
$$A = 2.828R^2$$
$$A = 3.314r^2$$
$$R = 1.307s$$
$$R = 1.082r$$
$$r = 1.207s$$
$$r = 0.924R$$
$$s = 0.765R$$
$$s = 0.828r$$

(b) Regular Octagon

Figure 9-10. Regular polygons.

VOLUME FORMULAS FOR SOLID FIGURES

Volume problems are often used to determine material costs, weights, treatments, and order quantities. There are several formulas that the apprentice mechanic should know for figuring the volumes (V) of common solid figures.

CUBE

A cube is a six-sided figure in which all the sides are squares of the same size. (See Figure 9-11.) The volume of a cube and its related dimensions can be calculated using the following formulas:

$$V = s^3$$

$$s = \sqrt[3]{V}$$

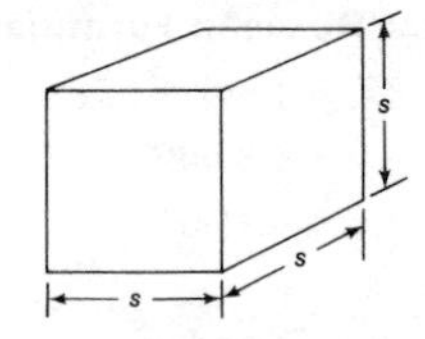

Figure 9-11. Cube.

PRISM

A prism is a solid figure that has equal polygons as ends and the ends are joined by parallelograms. There are two prism volume formulas that are very useful, one for calculating the volume of square prisms. (See Figure 9-12a.) and one for calculating the volume of any shaped prism. (See Figure 9-12b.)

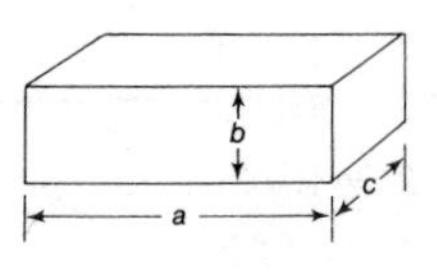

(a) Square Prism

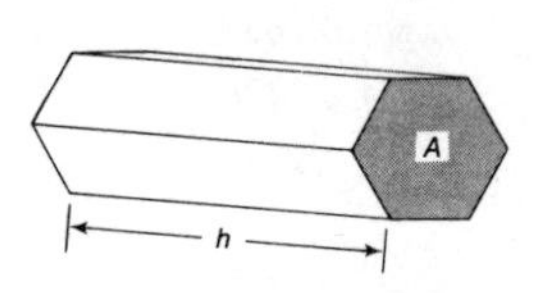

(b) Regular Prism

Figure 9-12. Prisms.

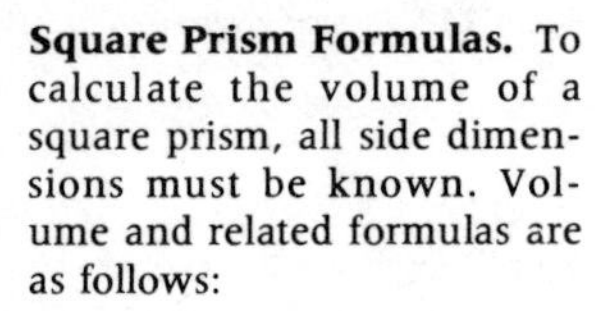

Square Prism Formulas. To calculate the volume of a square prism, all side dimensions must be known. Volume and related formulas are as follows:

$$V = abc$$

$$a = V/bc$$

$$b = V/ac$$

$$c = V/ab$$

Other Prism Formulas. Another formula can be used to calculate the volume of any type of polygonal prism. This formula requires that the area of the base be known or calculated, along with the length of the sides. The formula is:

$$V = hA$$

PYRAMID

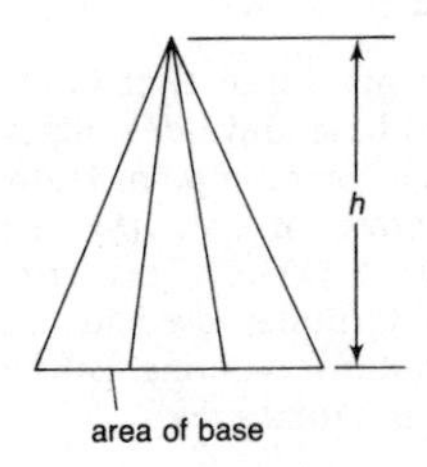

Figure 9-13. Pyramid.

A pyramid is a solid with a polygon base and triangular sides that meet a common point known as the *vertex*. (See Figure 9-13.). In the formulas used for volume calculations, n = the number of sides that are found in the regular polygon (base), s = the length of the sides, r = the radius of the inscribed circle, and R = the radius of the circumscribed circle. The formulas are:

$$V = ha/3$$
$$V = nsrh/6$$
$$V = (nsh/6)\sqrt{R^2 - s^2/4}$$

CYLINDER

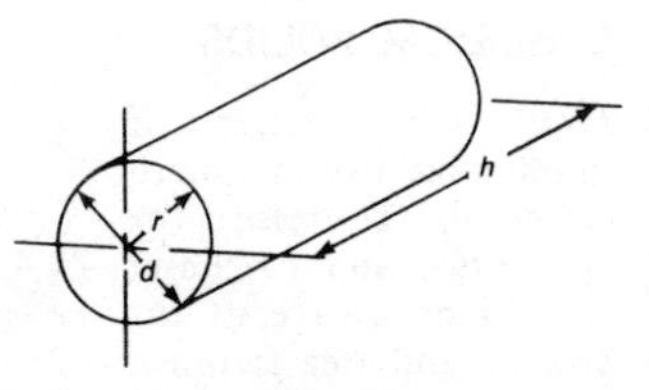

Figure 9-14. Cylinder.

A cylinder is a solid figure that has as ends circles of equal diameter and parallel to each other (see Figure 9-14). Formulas for the volume (V), side surface area (S), and total surface area (A) of a cylinder are used. Surface area refers to the area of the cylinder's bases and sides; this measurement is often important in determining material requirements. The total surface area of the cylinder is the area of the sides plus the end area. The formulas are:

$$V = \pi r^2 h$$
$$V = \pi d^2 h/4$$
$$S = 2\pi rh$$
$$S = \pi dh$$
$$A = 2\pi r(r + h)$$
$$A = \pi d[(d/2) + h]$$

CONE

Cones are solids that have a curved base and sides that taper to a common point known as a *vertex,* or *apex*. (See Figure 9-15.) There are three major formulas used in cone calculations: volume, surface area, and total area.

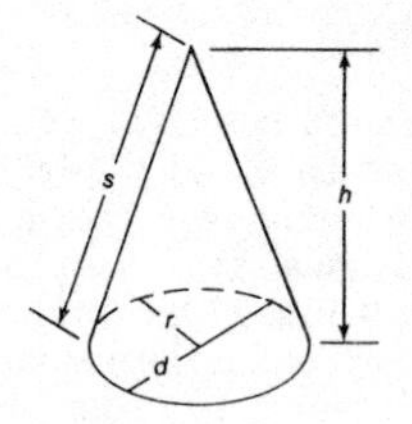

Figure 9-15. Cone.

$$V = r^2h/3$$
$$V = \pi d^2h/12$$
$$S = \sqrt{r^2 + h^2}$$
$$S = \sqrt{(d^2/4) + h^2}$$
$$A = \pi r\sqrt{r^2 + h^2}$$
$$A = \pi rs$$
$$A = \pi ds/2$$

SPHERICAL SOLIDS

Many storage tanks and units make use of spherical solids. Therefore the apprentice and mechanic should be aware of the volume and area formulas for these figures. (See Figure 9-16.)

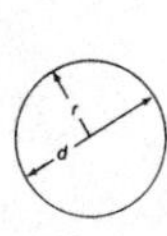

(a) Sphere

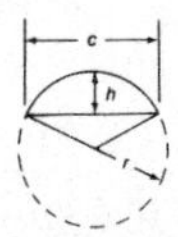

(b) Spherical Sector

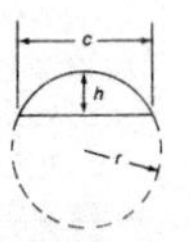

(c) Spherical Segment

Figure 9-16. Spherical solids.

Sphere. A sphere is a solid on which every point on the surface is the same distance from its center point—it is ball-shaped. Formulas used for calculating surface area and volume are as follows:

$$V = 4\pi r^3/3$$
$$V = \pi d^3/6$$
$$A = 4\pi r^2$$
$$A = \pi d^2$$

Spherical Sector. A spherical sector is that portion of a sphere that is defined by its radii and the spherical arc that is included within them. A spherical sector looks like a cone with a domed top. The basic formulas used are:

$$V = 2\pi r^2 h/3$$
$$A = \pi r[(2h) + c/2]$$
$$c = 2\sqrt{h(2r - h)}$$

Spherical Segment. A spherical segment is that section of a sphere that has to be cut away in order to leave a flat circular base. the volume and related formulas are as follows:

$$V = h^2[r - (h/3)]$$
$$V = \pi h[(c^2/8) + (h^2/6)]$$
$$A = 2\pi rh$$
$$A = \pi[(c^2/4) + h^2]$$
$$c = 2\sqrt{h(2r - h)}$$
$$r = (c^2 + 4h^2)/8h$$

MATHEMATICAL TABLES

Every mathematical handbook contains numerous tables useful for trade requirements. Two tables that are most valuable as ready references are decimal equivalency and functions of whole numbers. Table 9-1 lists fraction and decimal equivalents. Table 9-2 presents several functions of whole numbers, including squares, cubes, square roots, cube roots, and reciprocals of numbers from 1 to 50.

TABLE 9–1

Decimal Equivalents of Common Fractions

8th	16th	32nd	64th	Exact Decimal Values
			1	0.015625
		1	2	0.3125
			3	0.46875
	1	2	4	0.625

TABLE 9–1 (*continued*)

8th	16th	32nd	64th	Exact Decimal Values
			5	0.078125
		3	6	0.09375
			7	0.109375
1	2	4	8	0.125
			9	0.140625
		5	10	0.15625
			11	0.171875
	3	6	12	0.1875
			13	0.203125
		7	14	0.21875
			15	0.234375
2	4	8	16	0.25
			17	0.265625
		9	18	0.28125
			19	0.296875
	5	10	20	0.3125
			21	0.328125
		11	22	0.34375
			23	0.359375
3	6	12	24	0.375
			25	0.390625
		13	26	0.40625
			27	0.421875
	7	14	28	0.4375
			29	0.453125
		15	30	0.46875
			31	0.484375
4	8	16	32	0.50
			33	0.515625
		17	34	0.53125
			35	0.546875
	9	18	36	0.5625
			37	0.578125
		19	38	0.59375
			39	0.609375
5	10	20	40	0.625

TABLE 9–1 (*continued*)

8th	16th	32nd	64th	Exact Decimal Values
			41	0.640625
		21	42	0.65625
			43	0.671875
	11	22	44	0.6875
			45	0.703125
		23	46	0.71875
			47	0.734375
6	12	24	48	0.75
			49	0.765625
		25	50	0.78125
			51	0.796875
	13	26	52	0.8125
			53	0.828125
		27	54	0.84375
			55	0.859375
7	14	28	56	0.875
			57	0.890625
		29	58	0.90625
			59	0.921875
	15	30	60	0.9375
			61	0.953125
		31	62	0.96875
			63	0.984375
8	16	32	64	1

TABLE 9–2

Functions of Whole Numbers from 1 to 50

N Number	N^2 Square	N^3 Cube	$\sqrt{N}$ Square Root	$\sqrt[3]{N}$ Cube Root	$1/N$ Reciprocal
1	1	1	1.0000	1.0000	1.000000000
2	4	8	1.4142	1.2599	0.500000000
3	9	27	1.7321	1.4422	0.333333333
4	16	64	2.0000	1.5874	0.250000000
5	25	125	2.2361	1.7100	0.200000000

TABLE 9–2 (*continued*)

N Number	N^2 Square	N^3 Cube	$\sqrt{N}$ Square Root	$\sqrt[3]{N}$ Cube Root	$1/N$ Reciprocal
6	36	216	2.4495	1.8171	0.166666667
7	49	343	2.6458	1.9129	0.142857143
8	64	512	2.8284	2.0000	0.125000000
9	81	729	3.0000	2.0801	0.111111111
10	100	1,000	3.1623	2.1544	0.100000000
11	121	1,331	3.3166	2.2240	0.090909091
12	144	1,728	3.4641	2.2894	0.083333333
13	169	2,197	3.6056	2.3513	0.076923077
14	196	2,744	3.7417	2.4101	0.071428571
15	225	3,375	3.8730	2.4662	0.066666667
16	256	4,096	4.0000	2.5198	0.062500000
17	289	4,913	4.1231	2.5713	0.058823529
18	324	5,832	4.2426	2.6207	0.055555556
19	361	6,859	4.3589	2.6684	0.052631579
20	400	8,000	4.4721	2.7144	0.050000000
21	441	9,261	4.5826	2.7589	0.047619048
22	484	10,648	4.6904	2.8020	0.045454545
23	529	12,167	4.7958	2.8439	0.043478261
24	576	13,824	4.8990	2.8845	0.041666667
25	625	15,625	5.0000	2.9240	0.040000000
26	676	17,576	5.0990	2.9625	0.038461538
27	729	19,683	5.1962	3.0000	0.037037037
28	784	21,952	5.2915	3.0366	0.035714236
29	841	24,389	5.3852	3.0723	0.034482759
30	900	27,000	5.4772	3.1072	0.033333333
31	961	29,791	5.5678	3.1414	0.032258065
32	1,024	32,768	5.6569	3.1748	0.031250000
33	1,089	35,937	5.7446	3.2075	0.030303030
34	1,156	39,304	5.8310	3.2396	0.029411765
35	1,225	42,875	5.9161	3.2711	0.028571429
36	1,296	46,656	6.0000	3.3019	0.027777778
37	1,369	50,653	6.0828	3.3322	0.027027027
38	1,444	54,872	6.1644	3.3620	0.026315789
39	1,521	59,319	6.2450	3.3812	0.025641026
40	1,600	64,000	6.3246	3.4200	0.025000000

TABLE 9–2 (*continued*)

N Number	N^2 Square	N^3 Cube	$\sqrt{N}$ Square Root	$\sqrt[3]{N}$ Cube Root	1/*N* Reciprocal
41	1,681	68,921	6.4031	3.4482	0.024390244
42	1,764	74,088	6.4807	3.4760	0.023809524
43	1,849	79,507	6.5574	3.5034	0.023255814
44	1,936	85,184	6.6332	3.5303	0.022727273
45	2,025	91,125	6.7082	3,5569	0.022222222
46	2,116	97,336	6.7823	3.5830	0.021739130
47	2,209	103,823	6.8557	3.6088	0.021276600
48	2,304	110,592	6.9282	3.6342	0.020833333
49	2,401	117,649	7.0000	3.6593	0.020408163
50	2,500	125,000	7.0711	3.6840	0.020000000

10

Blueprint Reading

Machine and maintenance mechanics are frequently required to read and interpret technical and engineering drawings of facilities, machinery, and equipment. These drawings are used for the installation, maintenance, and repair of equipment. In addition, architectural drawings are used for facility layout and location requirements. This chapter presents a basic overview of how to read and interpret these prints.

DRAWING PRESENTATIONS

The term "blueprint" is often used within the trade to identify any type of reproduced drawing. Technically, the term refers to a specific type of reproduction in which the print is made by exposing presensitized paper to ultraviolet light while the original drawing is placed on top of it. After exposure, the paper is sent through a series of baths that result in a print in which the lines are white and the background is a deep blue—a print that appears like a photographic negative.

Today, most prints are made by an ammonia vapor process known as diazo prints. The original is again placed over presensitized paper and exposed to ultraviolet light. The paper is then developed in an ammonia vapor, producing a blue or black line print with a white background—a positive print. In most cases, the mechanic will be working with this type of print—which is still referred to as a blueprint in many sectors.

ORTHOGRAPHIC PROJECTIONS

All technical and engineering drawings are based on the principles of orthographic projection, a technique or method for drawing various views of an object whereby one can accurately interpret how that object appears in three dimensions. In most drawings, there are three primary views of projection: front, top, and right side (profile).

The views in drawings show how the object would appear if viewed through imaginary planes positioned about it. Figure 10-1 is a comparison between a three-dimensional object and its frontal orthographic view. In orthographic projection, the lines of projection are drawn perpendicular to the plane so that all dimensions are true size. The word "orthographic" is derived from the Greek *orthos,* meaning straight or right angle, and *graphic,* meaning drawing. Hence, orthographic projection is a drawing technique using straight and perpendicularly projected lines.

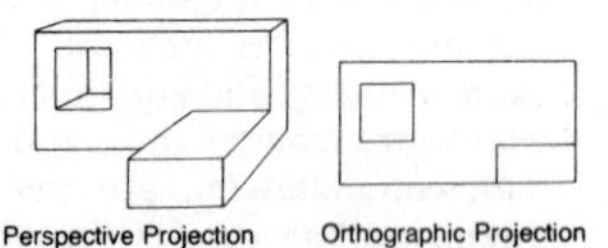

Figure 10-1. A comparison between a three-dimensional view (perspective projection) and a frontal view (orthographic projection).

SIX PRINCIPAL VIEWS OF PROJECTION

If an object is enclosed in a glass box in which each side is perpendicular to the other, the object could be viewed through six different planes (or sides). If the object were drawn on each side of the box as viewed through that glass plane, the projections would show the front, top, right side, left side, bottom, and rear views of the object. These six views are known as the six principal views of projection. (See Figure 10-2.)

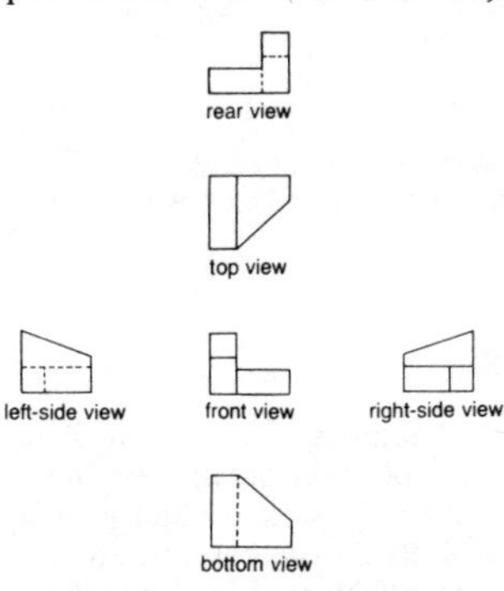

Figure 10-2. Six principal views of projection.

Though it is possible to show six principal views of an object, most drawings only show three primary views: front, top, and right side. Only in unusual or ambiguous situations will other views be shown. When a drawing is made with the three primary views, it is important that they be properly positioned on the drawing sheet. Thus, the top view should be directly above and aligned with the front view. In turn, the right side view should be directly to the right and aligned with the front view.

TRUE LENGTH LINES

An important point that should be understood by anyone who works with drawings is true length lines, which are lines that appear true size in a given view. To be true length, the line has to meet one criteria: it must be parallel to the viewing plane. When shown in its true length, however, it is not necessarily parallel to any other view.

FRONT VIEW

In the drawing process itself, the drafter will decide which surface of the object will be used as the front view. The basic rule-of-thumb is that the single view that would best describe an object should be used as the front view. (See Figure 10-3.) Drawing procedures will dictate that the front view be drawn first. Light projection lines are then drawn to aid the drafter in locating the top and side views.

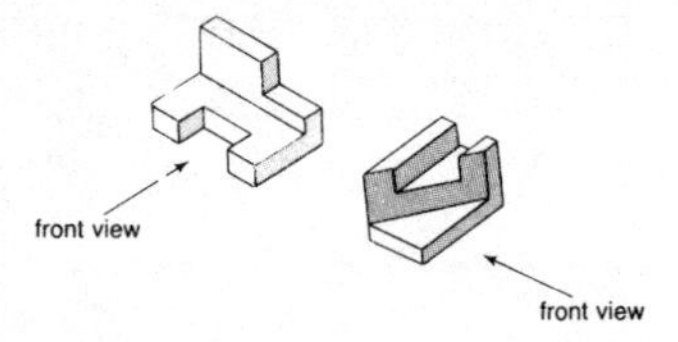

Figure 10-3. The view that best describes an object should be used as the front view.

VIEWS AND MEASUREMENTS

By technical definition, the front (frontal) view will only show length and height measurements. The top (horizontal) view shows length and width measures, while the right side (profile) view illustrates width and height measures. Hence, within a three-view drawing, one will be able to see a given feature at least two times. Within the machine trades, especially those incorporating numerical control and computerized systems, measurements are referred to by axis coordinates, defined as:

X axis: length
Y axis: width or depth
Z axis: height

Trying to Visualize a Three-View Drawing

Step 1 Study the entire drawing, first noting the number of views and their placement.

Step 2 Look at each view and decide which one best describes the object—this should be the front view.

Step 3 Observe and study the top and side views for lines that show the boundaries of surfaces. This is usually defined by the intersection of lines.

Step 4 Carefully study each view at one time, taking into consideration the overall shape of the view, its surfaces, lines, and planes.

Step 5 If necessary, do a sketch to help you visualize what the object looks like three dimensionally. Make sure that you give close attention to the location and shape of each feature on the object.

When working with multiple view drawings remember that no one view completely describes the object. This brings up an important point: it is not always necessary to show three views of an object. This is particularly true when showing symmetrical objects. (See Figure 10-4.) In this example, only two views are needed since the top view would be exactly the same as the front view. Because only two views are needed to adequately describe the object, no more are necessary.

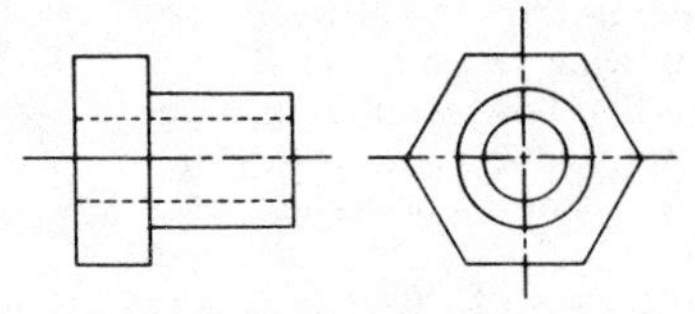

Figure 10-4. It is not necessary to show three views of symmetrical objects.

DETAIL AND ASSEMBLY DRAWINGS

Generally, industrial drawings will be either detail or assembly drawings. Detail drawings are those that illustrate one product or part and provide all necessary information and specification for its manufacture. Assembly drawings show multiple-part products as they would appear totally assembled. Drawings that are referred to as subassembly drawings are sometimes used for very complex products. In these presentations, component assemblies of the total assembly are presented.

PICTORIAL DRAWINGS

Not every drawing encountered will use orthographic projection techniques. In a few cases, pictorial drawings are used to illustrate assembly or installation procedures. The pictorial drawing is the oldest form of graphic communication. There are three broad categories of pictorial drawings: axonometric, oblique, and perspective. Of these, only the first two are usually encountered by the mechanic.

AXONOMETRIC PROJECTION

The principles involved in axonometric projection are similar to orthographic projection, except that the view illustrated shows the object tilted in such a manner that it gives a sense of perspective or depth. The most common form of axonometric projection used is the isometric drawing.

Isometric Drawing. When drawing an isometric, the object is shown so that its edges (axes) form equal angles with the plane of projection. (See Figure 10-5.) As illustrated, the axes of the object form three equal angles of 120° (30° from the horizontal) with the proportion of all edges (height, length, and width) being the same.

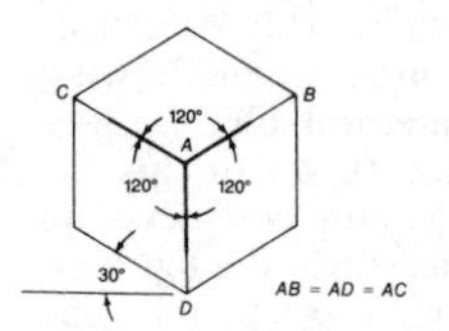

Figure 10-5. Isometric cube.

OBLIQUE PROJECTION

Oblique projection drawings are used when one surface of an object is of critical importance. In essence, one surface of the object is drawn as it would appear in a typical orthographic projection view—appearing true size and shape. From this surface, other surfaces are drawn away from it at given angles, in order to present a pictorial presentation. (See Figure 10-6.)

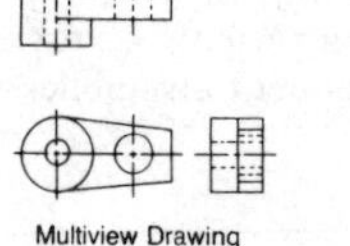

Figure 10-6. Oblique projections.

As in axonometric drawings, the angle and proportions of each axis will sometimes vary in an oblique drawing. Typical angles of inclination are 30°, 45°, and 60°. When the receding axis's scale is drawn full size, the drawing is known as a *cavalier oblique*. Drawings with a half size receding axis are *cabinet obliques.* Cavalier and cabinet oblique are the two most common forms of scaling drawings. However, other foreshortening proportions are sometimes used to present a more realistic or understandable presentation.

DRAWING CONVENTIONS

The interpretation and understanding of drawings is an essential skill for all mechanics. Thus it is important for the mechanic to understand the drawing conventions used within the trade.

The vast majority of drawing standards used in the United States are also used internationally. There are, however, some international standards not commonly used in the United States. Since many pieces of equipment and machinery are made by international firms, it is important for mechanics to be able to read drawings that are made according to either national or international standards.

National standards are provided by the American National Standards Institute (ANSI); international standards are provided by the International Organization for Standardization (ISO). When necessary, the distinction between ANSI and ISO practices will be pointed out.

LINE EXPRESSIONS

All drawings are made with lines whose individual characteristics define and communicate specific information. Within the industrial societies of the world, specific standards have been developed so that a drawing made in New York or California can be interpreted by a mechanic in Germany or Australia. The set of conventional line expressions used in technical and engineering drawings is often referred to as the *alphabet of lines.*

There are eight common line expressions used in drawings. (See Figure 10-7.) All lines are drawn black. What differentiates one line

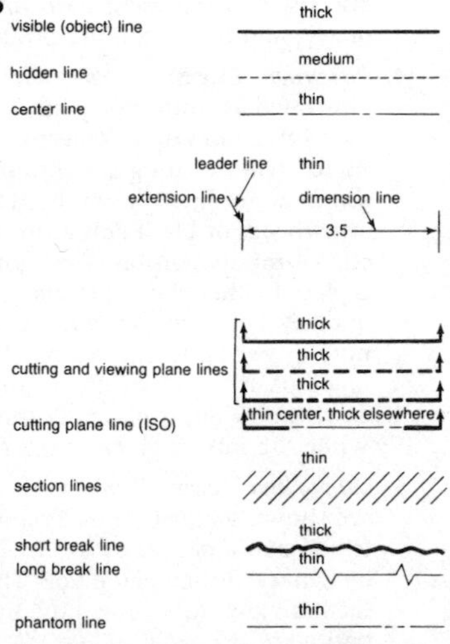

Figure 10-7. The alphabet of lines.

from another is its thickness and the way it is drawn. The characteristics and uses of these lines are described in the alphabet of lines list.

Alphabet of Lines

1. *Visible Lines.* Visible lines are shown as continuous thick lines. They are used to represent visible outlines and edges of objects. Sometimes called *object lines*, they represent only those portions of an object that can be seen in a particular view.
2. *Hidden Lines.* Hidden lines are drawn as short thin dashes. They are used to show hidden outlines and edges. In some cases, a view may not show all hidden lines if they would complicate or confuse the reader. When these lines do appear, refer to an adjoining view for further clarification and description.
3. *Center Lines.* Drawn as thin chain lines with a long and short dash pattern, center lines are used to locate the center of round or symmetrical shapes or objects. In some situations they are also used to represent path lines, pitch circles, axes of symmetry, and the maximum position of movable parts.
4. *Dimension, Extension, and Leader Lines.* These types of lines are illustrated as continuous thin lines. *Extension lines* extend from the visible lines of an object to indicate the limits of a dimensioned feature; they are continuations of object lines. *Dimension lines* are lines drawn between extension lines; they have arrowheads or black dots at the ends that indicate the start and end of the dimension. The numerical value of the dimension is placed either above the dimension line, or in an opening, left midway between the extension lines. *Leader lines* are used to note a specific feature or part. They are usually drawn at an angle of 30°, 45°, or 60° to the center of the feature. An arrowhead or dot is placed at the feature end of the leader line, while the note is given at the other end.
5. *Cutting and Viewing Plane Lines.* Cutting and viewing plane lines are shown as thick lines. Primarily used to show the internal features of a part, they are used to indicate the position of an imaginary cutting plane line. These cutting planes are used for section views that show how the object would appear if a section were cut away.

6. *Section Lines.* Drawn on a slant as thin continuous lines, section lines are used in section views to indicate the surfaces that have been cut away.
7. *Break Lines.* Break lines are drawn where a part is too long for a drawing or view and must be foreshortened. There are three types of break lines: the first is a thick wavy line used for short breaks; the second is a thin straight line with a zigzag, or "Z," drawn in the middle for a long break; the last is an "S" break used for cylindrical objects.
8. *Phantom Lines.* Drawn thin with a long line followed by two short dashes, phantom lines are used to indicate moving parts in extreme positions and their path. They are also used to outline adjacent parts, the initial outline of the object prior to final forming or finishing, as well as parts that are located in front of the cutting plane line.

LETTERING

In addition to the graphic representation of objects, drawings also provide information through letters and numbers. Most lettering on drawings is done by hand, although mechanical devices such as templates, lettering machines, and press-on letters may also be used.

Most lettering will be single-stroke vertical Gothic. The size of the lettering depends on the information to be provided. Most drawing dimensions, notations, and specifications will be 1/8 inch (3mm) in height. Lettering between 3/16 and 1/4 inch (5-6 mm) is used for headings; lettering up to 1/2 inch (13 mm) is used for titles and title blocks.

DIMENSIONING

Two basic dimensions are given on drawings: size and location dimensions. Size dimensions tell the print reader the size of a part or feature, such as a hole, arch, fillet, flat surface, or arc. Location dimensions are used to determine the position of a given feature relative to another feature such as an edge or hole.

To ensure print reading accuracy and efficiency, a mechanic should be familiar with the basic dimensioning principles followed on all drawings.

Basic Dimensioning Principles

1. All dimensions that are required for the part to function correctly are given directly on the drawing. In addition, dimensions needed to describe the part completely in its finished form should be shown; this includes any manufacturing, inspection, or servicing requirements.
2. The same dimension is not repeated in a drawing unless unavoidable. The dimension will appear in the view that shows the relevant feature or part most clearly.
3. All dimensions are given in the same unit, such as inches or millimeters. If not, a special notation will be made.
4. Only those dimensions needed to describe the end product or feature locations are given. The only exceptions involve special cases in which the dimension is needed for intermediate manufacturing or installation processes or it is desirable to add further dimensioning.
5. Features that play an essential part in the performance or serviceability of the product (i.e., functional features) are dimensioned directly on the drawing. Nonfunctional features are dimensioned in any convenient manner.
6. Tolerances are specified for all requirements influencing function, operation, or interchangeability.
7. Standard sizes are used wherever possible and practical.
8. Only essential manufacturing, assembly, inspection, and servicing methods are specified on drawings.

DIMENSIONING SYSTEMS AND PRACTICES

There are several acceptable dimensioning systems used alone or in combination in drawings.

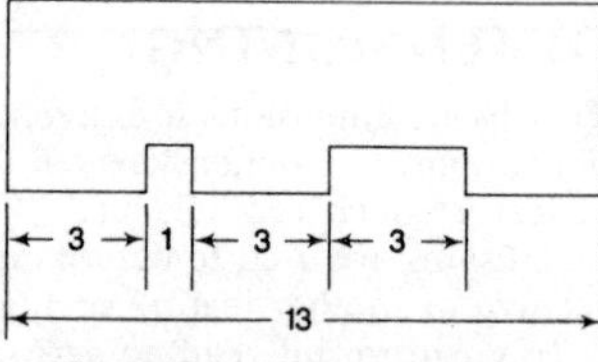

Figure 10-8. Chain dimensioning.

Chain Dimensioning. The first, known as chain dimensioning, is used in those cases in which an accumulation of dimensional error will not have an effect on the functional requirements of the part. As the name implies, each dimension starts with a preceding dimension—a chain effect. (See Figure 10-8.)

Parallel Dimensioning. A second dimensioning system is known as parallel dimensioning. Here, all dimensions have a common datum, or reference point, from which all measures are taken. There are several ways in which this system can be used. (See Figure 10-9.) The conventional method uses a datum extension line and series of dimension lines where the shortest dimension is closest to the object, and the largest is farthest away. Where advantageous, a simpler procedure can be used. The datum line is noted by a dot and a zero. The dimensions are then lettered in at each extension line. Note that arrowheads are placed on one side of the dimension line to indicate its distance from the datum point. However, care must be taken to avoid any confusion.

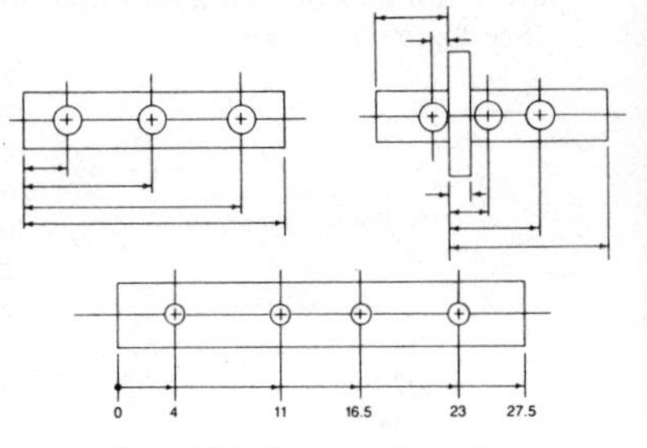

Figure 10-9. Examples of parallel dimensioning.

Coordinate Dimensioning. The last dimensioning system to be discussed, known as coordinate dimensioning, has found great acceptance in numerical control and computer-aided machining. Here, instead of dimensioning a part by chain or parallel methods, specific dimensions may be grouped and identified by coordinate positions along axes. Figure 10-10 gives an example of this system, where the location of each hole is given according to its location along the X and Y axes.

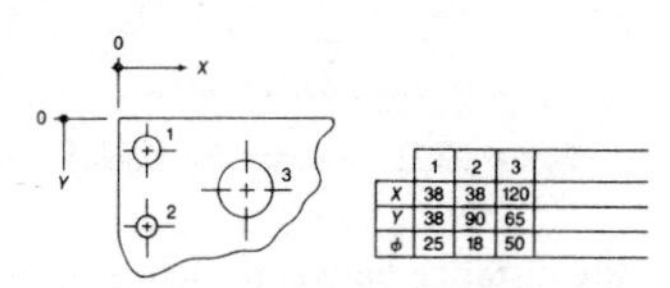

Figure 10-10. Coordinate dimensioning.

SPECIAL DIMENSIONING PRACTICES

In addition to the standard dimensioning practices discussed, there are a number of techniques that are used for special situations. (See Figure 10-11a.)

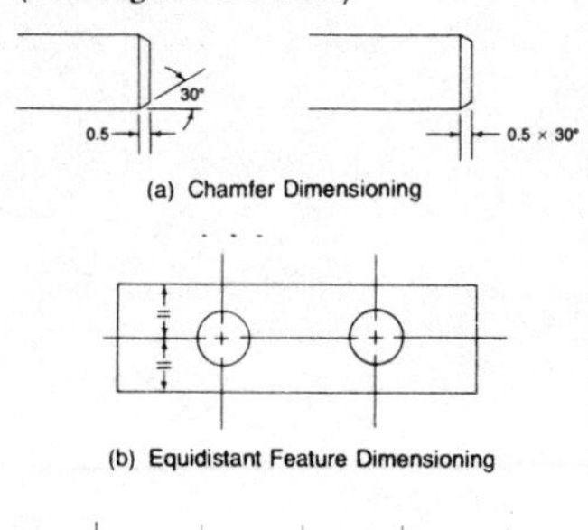

(a) Chamfer Dimensioning

(b) Equidistant Feature Dimensioning

(c) Repeated Equidistant Feature Dimensioning

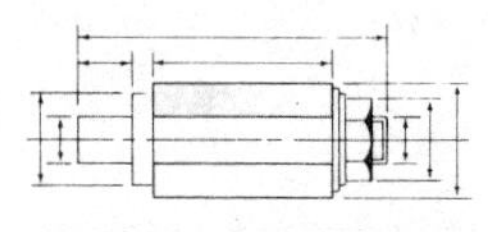

(d) Dimensioning Assembled Parts

Figure 10-11. Special dimensioning practices.

The first technique involves the dimensioning of chamfers. There are two basic methods available here. The first is to give the width of the chamfer and its angle. The second is a notation procedure whereby the extension lines project from the chamfer's width and are dimensioned as width *x* angle.

Another special practice involves the dimensioning of equal distances. When the dimension is divided into a given number of equal parts, the equals sign (=) may be used to show that those dimensions are nominally equal to each other. (See Figure 10-11b.)

There are a number of products made with features that are equidistant from one another where an abbreviated dimension form may be used. As shown in Figure 10-11c, the distance between each feature is multiplied by the number of features and equaled to the total distance from start to end. If there is a potential for confusion between the distance and the number of features, one of the features may be fully dimensioned.

The last special practice to be considered is the dimensioning of assembled parts. (See Figure 10-11d.) This is particularly common when complete units are received for servicing and/or installation. When a number of different parts are drawn in assembly, the dimensions given for each part will be kept separate as much as possible.

NOTATIONS AND SYMBOLS

Prints used within the trade often include special types of notations, the nature of which depends on the type of drawing.

NOTATIONS FOR TOLERANCING

One of the most important notation procedures used today is that of tolerancing. A tolerance is expressed in the same form as a dimension. For example, a fractional dimension would have a fractional tolerance such as: 6 5/8 ± 1/32. A decimal tolerance would have a decimal tolerance: 2.750 ± 0.003. Angular tolerances are given in terms of degrees (°), minutes ('), or seconds ("): 45° ± 15'.

Within the area of precision machines and instruments, specific surface tolerances are noted by the use of other special symbols that identify the nature of tolerancing. (See Figure 10-12.) When given on drawings, tolerance characteristics are often accompanied by additional tolerancing information.

	Characteristics to be toleranced	Symbols
Form of single features	Straightness	—
	Flatness	⏥
	Circularity (Roundness)	○
	Cylindricity	⌭
	Profile of any line	⌒
	Profile of any surface	⌓
Orientation of related features	Parallelism	//
	Perpendicularity (Squareness)	⊥
	Angularity	∠
Position of related features	Position	⊕
	Concentricity and coaxiality	◎
	Symmetry	⌯
Run-out		↗

Figure 10-12. Tolerance characteristic symbols.

Rectangular Frame. Additional information is frequently presented in a rectangular frame, which contains three basic compartments. From left to right, the following information is given:

1. The symbol for the tolerance characteristic, as shown in Figure 10-12.
2. The total value of the tolerance in the unit used for drawing linear dimensions. This value may be preceded by other symbols, such as those noting circular, cylindrical, or square features.
3. Sometimes it is necessary to identify a given datum feature or features. Here, a letter or letters will be used.

Examples of the rectangular frames are:

//	0.2	A	0.1

When dimensioning with rectangular frames, the tolerance frame is connected to the part's feature by a leader line with an arrow at one end. Datum features that are toleranced are indicated by a leader line that ends in a solid triangle, whose base lies on that feature. Examples of this are illustrated in Figure 10-13.

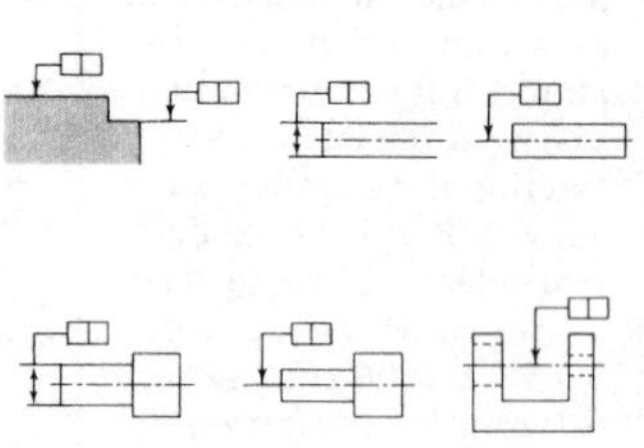

Figure 10-13. Examples of how tolerance frames are used in dimensioning.

Surface Texture Indications. Another tolerancing method is an indication of surface texture. This method is limited to precision machining operations used in manufacturing processes. (See Figure 10-14.)

1. *Roughness.* Roughness refers to the finely spaced irregularities of the finished surface that are produced by machining operations. This feature is expressed in microinches (0.00001 in.)
2. *Waviness.* Waviness refers to surface irregularities that are spaced too far apart to be considered roughness. This feature

is usually caused by factors such as machine and work deflection, vibration, chatter, heat treatment, and strain warpage.

3. *Lay.* Lay is the primary direction in which the surface irregularity appears. It is sometimes referred to as tool markings. This characteristic is represented by several graphic symbols. (See Figure 10-15.)

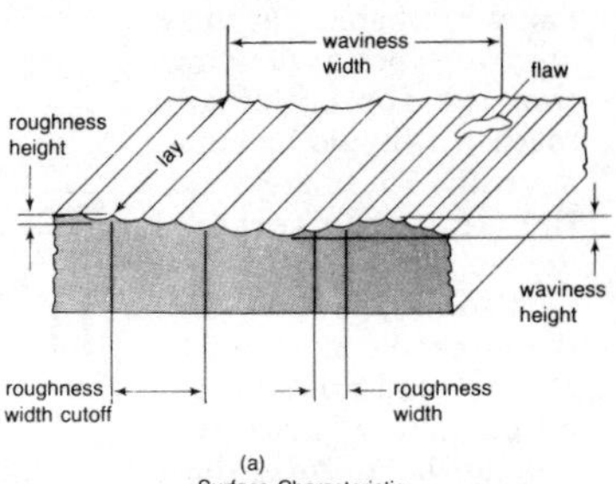

Title Block. Most tolerancing information is presented in the title block of the drawings. The methods just described are used for specialized features. Examples of several title block notations are:

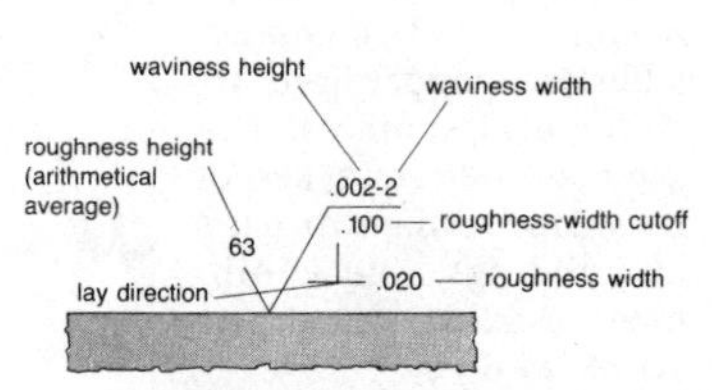

Figure 10-14. Surface texture identification and notations.

TOLERANCES UNLESS OTHERWISE SPECIFIED:
FRACTIONS ±1/64
DECIMALS ±0.003
ANGLES ±30′

NORMAL TOLERANCES:
FRACTIONAL DIMENSIONS ±.015
DECIMAL DIMENSIONS ±.003
ANGULAR DIMENSIONS ±1 DEGREE

±.020 TOLERANCE ON TWO PLACE DECIMALS
±.010 TOLERANCE ON THREE PLACE DECIMALS
ZERO TOLERANCE ON FOUR PLACE DECIMALS
±30′ ANGLE TOLERANCE

SYMBOLS FOR SPECIAL DRAWINGS

There are a number of symbols that are used in special drawings. Not all are commonly encountered by the machine and maintenance mechanic, but there are two types of drawings that the apprentice and mechanic should be familiar with. These are welding and architectural drawings.

In some job situations, the mechanic must also be able to read prints in additional trade areas. It is not possible to cover these topics in this book.

Architectural Symbols. The American Institute of Architects (AIA) has established standardized architectural symbols. The symbols indicate types of materials, sections or parts of buildings, and equipment installed. (See Figure 10-16, a–d.)

Welding Symbols. Welding symbols are used in a variety of fabrication and repair drawings. In these drawings, welding information is given by use of a weld symbol. (See Figure 10-17.) The arrow in the symbol points to the

Surface Texture Lay Symbols

Lay Symbol	Designation	Example
‖	lay parallel to the line representing the surface to which the symbol is applied.	direction of tool marks
⊥	lay perpendicular to the line representing the surface to which the symbol is applied.	direction of tool marks
X	lay angular in both directions to line representing the surface to which symbol is applied.	direction of tool marks
M	lay multidirectional.	
C	lay approximately circular relative to the center of the surface to which the symbol is applied.	
R	lay approximately radial-relative to the center of the surface to which the symbol is applied.	

Figure 10-15. Surface texture lay symbols.

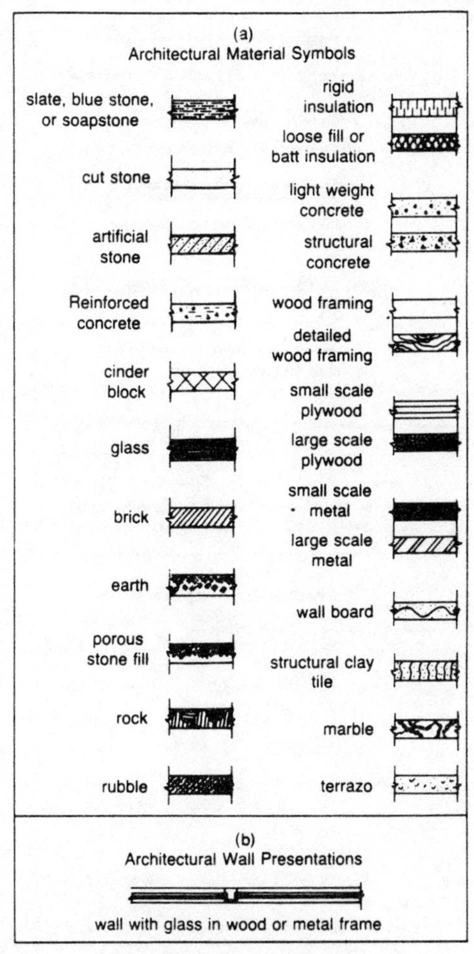

Figure 10-16. Architectural symbols.

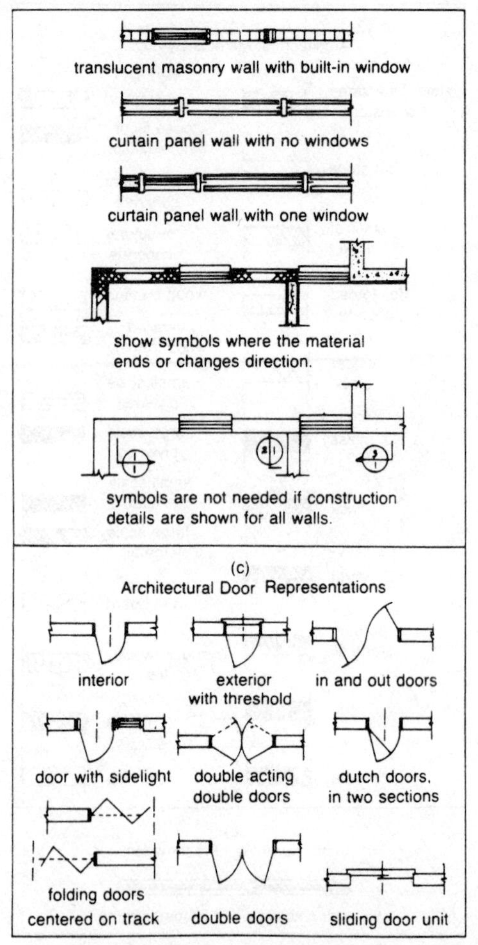

Figure 10-16. (continued)

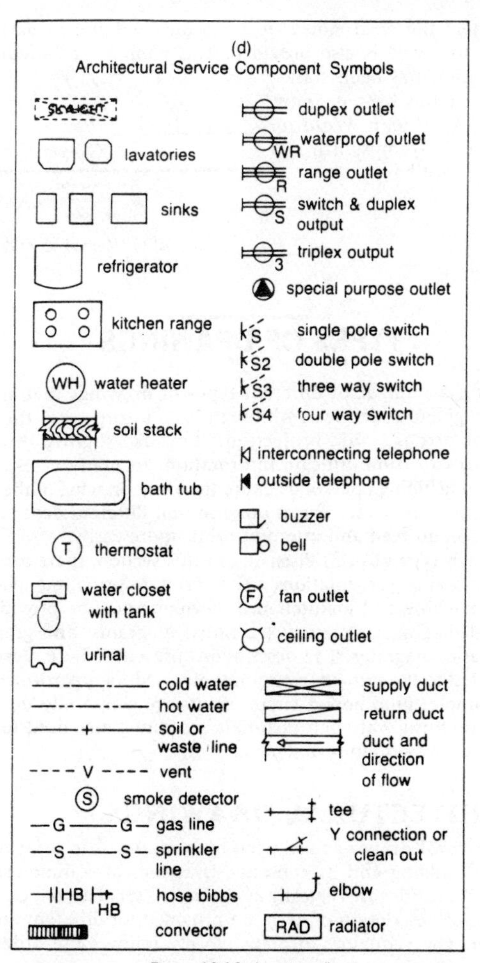

Figure 10-16. (continued)

spot where the weld joint is to be located. Additional data concerning the weld is also provided. The symbols and method of presentation has been standardized in the United States by the American Welding Society (AWS), on which ISO standards are based.

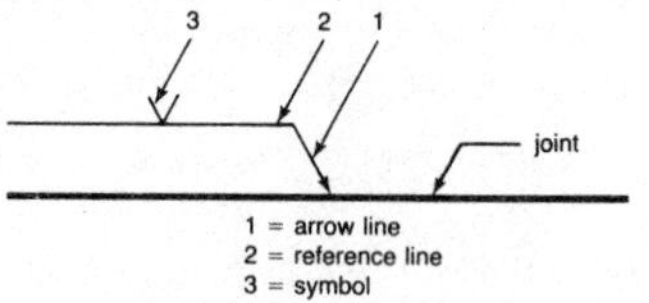

Figure 10-17. Method of welding representation.

TYPES OF DRAWINGS

There are a number of different types of drawings that are peculiar to given trade areas. Although each incorporates the principles of orthographic projection, they use slightly different procedures to communicate information. In many cases, especially in machining operations, more than one drawing of the same product may be made. This is done so that different departments will be able to read and interpret prints more easily.

For each type of print discussed in this section, there are additional drawing presentations used. Print information available to the machine and maintenance mechanic may be provided by pictorial diagrams, cutaway (sections) diagrams, and graphical (schematic) diagrams. The descriptions presented here, however, are limited to the prints most commonly used for a particular area. For example, fluid power circuit prints are commonly drawn in schematic form, although pictorials are sometimes used to illustrate how the system is designed.

ARCHITECTURAL DRAWINGS

Architectural drawings are used to communicate information about a building and its contents. Drawings of commercial and industrial facilities are typically available as a set of drawings. Plans, and a brief description of each are provided for different areas of concern. The sequence of plans closely follows the order that would appear in a set.

SITE PLANS

Site plans are drawings that show the relationship between the building and the natural features of the property or site on which it is built. In these plans, only a foot print (outline) of the building is provided while site details are elaborated.

1. Boundaries of the property, streets and alleys, their widths, names, and intended dedications to public use
2. Lines of all adjoining properties, streets, and alleys, with their widths and names
3. All lot lines and numbers, building lines, and utility easements
4. All angular and linear dimensions necessary for location and boundaries of the site
5. Contours at a standard vertical interval
6. Location of water mains, sewerage lines, and all utilities.

FOUNDATION PLANS

Foundation plans provide information about the substructures of the building. (See Figure 10-18.) These drawings have information about the earth and soil upon which the building is constructed, footing design and locations, and any walls built upon the footings.

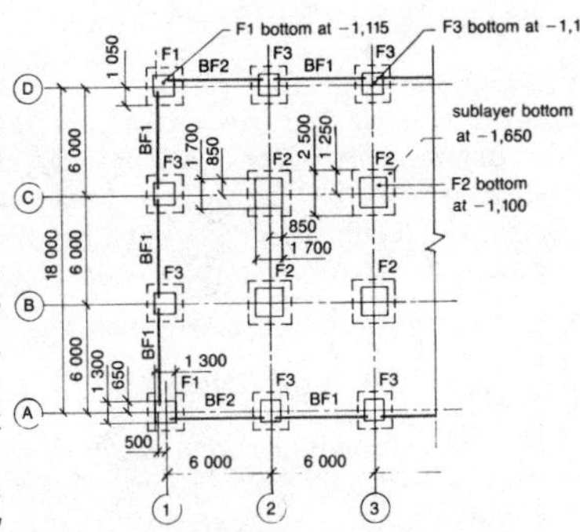

Figure 10-18. Typical foundation plan.

Associated with foundations plans are *structural plans*, which illustrate in detail the integration of the structural components of the system. These include footings and walls; column marks; footing, wall, and column specifications; and details showing the structural makeup of each.

FLOOR PLANS

Floor plans are drawings of a building as it would be viewed from above if its walls were horizontally cut and removed. Theoretically, floor plans show how a building would look if it were sectioned or cut off at a height of about four feet above the floor level. Architectural symbols for items such as type of wall material, windows, doors, electrical outlets, plumbing fixtures, and mill work are used extensively. (See Figure 10-19.)

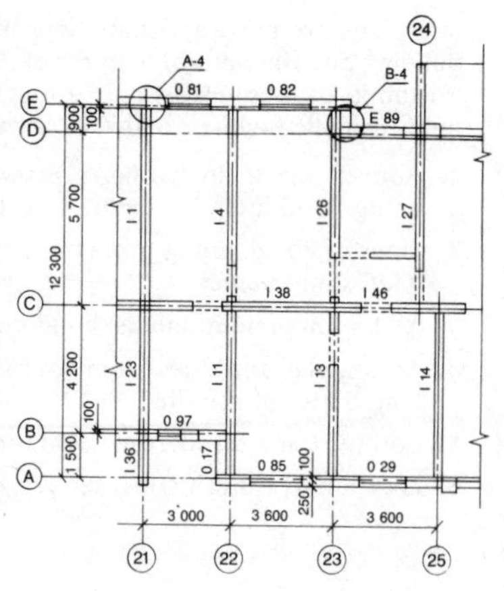

Figure 10-19. Section of a typical floor plan.

DETAIL DRAWINGS

Detail drawings are used to show special features and characteristics that are not clearly understood in other drawings. Details show how special construction or installation is to be accomplished. The number and types of details are determined by the nature of the building and how it differs from standard construction practices.

ELEVATION DRAWINGS

Elevation drawings show the building as it would appear in the frontal, back, and side views. There are two basic types of elevations: interior and exterior. Interior elevations are only used to clarify some construction or design consideration within the building itself. Exterior elevations show how the building appears from the outside. Both drawings are important because they provide all height measurements.

STRUCTURAL DRAWINGS

Structural drawings show the complete structural configuration of the building. In other words, it is the skeletal system of the building, showing all beams and supports. Important here are specifications relative to load bearings, structural dynamics, and stress points.

PLUMBING SYSTEM DRAWINGS

Plumbing system drawings show the design, layout, and component parts of the building's plumbing system. These drawings not only show water supply lines but also soil and waste system diagrams and other piping specifications. Most of these drawings are schematics that use standardized symbols.

ELECTRICAL SYSTEM DRAWINGS

Electrical system drawings show the layout and position of all electrical components. Like plumbing system drawings, electrical drawings are primarily schematics. The exception to this are power drawings that show the wiring details of units such as electrical panels.

HVAC DRAWINGS

HVAC drawings include information about the heating, ventilation, and air-conditioning systems in the building. They provide information about the type of heating and cooling units to be used in the system, heating and cooling ductwork location and capacities, location and size of fittings, turns, grills, and other aspects of the system.

ELECTRICAL PRINTS

Industrial facilities, especially those in which large electrical drive motors are used, make use of some of the most complex prints.

A typical electrical diagram is illustrated in Figure 10-20. This is a schematic diagram for a typical AC power switch gear. In the schematic diagram, component and connection information is given for the entire system, power supply values are provided, and individual components are alphanumerically coded.

Electrical prints such as the one shown may be difficult to interpret and understand without adequate training in the field. However, the apprentice and mechanic should be able to recognize such prints and be aware of the type of information included. Such prints are not only useful for initial wiring and installation but they are also helpful to electricians for troubleshooting and repairing equipment.

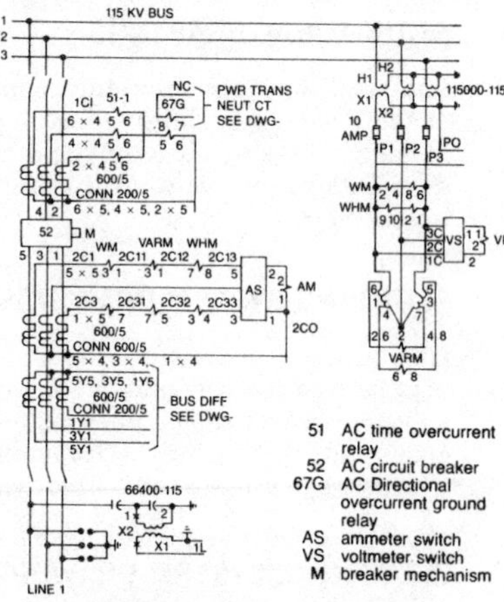

Figure 10-20. Example of typical electrical diagram for AC power switch gear.

FLUID POWER CIRCUIT PRINTS

A fluid power circuit is a control system that has recently been introduced in much machinery. Basically, a fluid power circuit system employs compressed air and hydraulic oils to regulate various machine components. Air and hydraulic systems are similar, though air systems have the advantage of requiring smaller units and components.

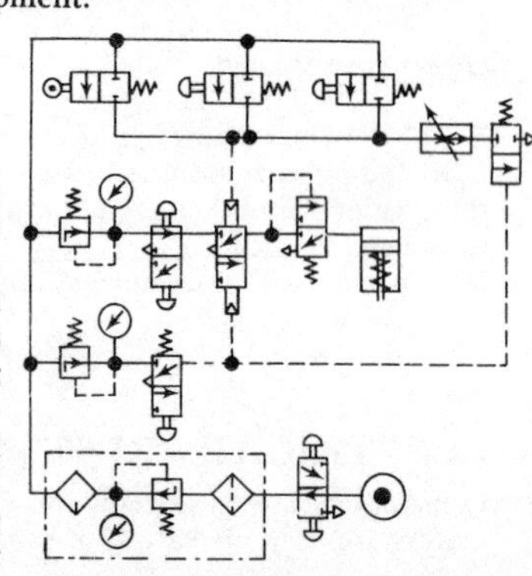

Figure 10-21. Typical fluid power diagram for a clutch operating control.

Fluid power circuit control

devices can accomplish the same functions found in electrical circuits. Figure 10-21 is an example of a complete clutch operating control. In these diagrams, symbols are used to identify equipment in an nonoperating condition. If some other condition is to be represented, it is clearly stated.

PIPE DRAWINGS

Piping drawings are presented in two common forms: layout and isometric. (See Figure 10-22.) These drawings are used for water supply systems, and they also incorporate hot water heating systems and low and high pressure gas systems. Pipe layouts are drawn to scale over facility floor plans. Isometric drawings, by comparison, are drawn free standing and are usually not drawn to scale; their primary purpose is to illustrate the general configuration of the system.

Again, specific graphic symbols (see architectural drawings) are used to illustrate piping components. Examples of piping systems requiring such drawings are wet pipe systems, dry pipe systems, pre-action systems, deluge systems, combined dry pipe and pre-action sprinkler systems, limited water supply systems, foam systems, CO_2 systems, and hose standpipe systems.

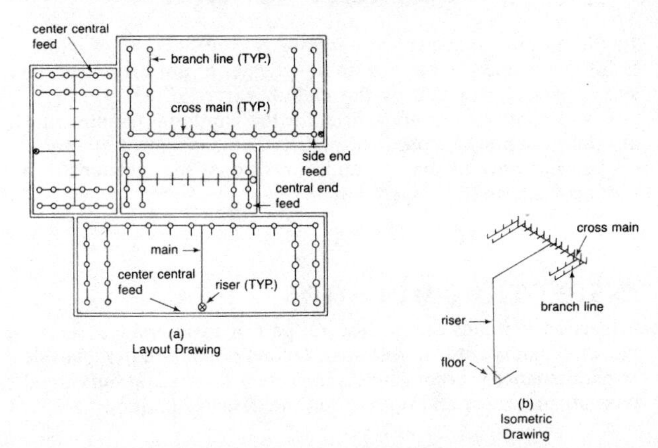

Figure 10-22. Typical piping drawings.

11

Measurement and Conversions

The manufacturing of machinery, equipment, and instruments is based on the interchangeability of parts. These parts must be accurately made, installed, and maintained to within various degrees of tolerance. Central to this are accurate measurement and alignment. This chapter presents an overview of the concepts involved in measurement and alignment as well as the conversion factors used for U.S. Customary and SI metric units.

INSPECTION AND QUALITY CONTROL

To the untrained, inspection and quality control is the examination of units and parts to determine whether or not they meet the size specifications given by the manufacturer. In reality, inspection and quality control are used for the continual maintenance and improvement of product quality. The concept of quality is applied not only to the size and dimensional specifications of a product but also to its operation.

INSPECTION METHODS

Inspection methods are an integral part of maintenance, servicing, and quality control. Although not every method is applicable to maintenance procedures, it is important for the apprentice and mechanic to know and understand the basic techniques.

DESTRUCTIVE INSPECTION

Destructive inspection is usually limited to the testing of the internal structure of a product's material. It is used to find out its homogeneity and resistance to stresses and strains that the product may be subjected to during service. Because destructive inspection renders the material useless for further application, this procedure is primarily employed during product manufacture.

DIMENSIONAL INSPECTION

Dimensional inspection is the examination of a product's size to ensure that it will function properly as a component part. In many cases, dimensional inspections are used as a go– no-go procedure to make sure that the part selected will fit properly and work.

NONDESTRUCTIVE INSPECTION

In nondestructive inspection a product's integrity is examined and tested without the product being destroyed. The major advantage of this procedure is that it allows for material testing without costly product destruction. Examples of these inspection techniques include X rays, magnaflux, electronic, laser beam, transducer tracer and indicator, and optical flat procedures.

SUPERFICIAL INSPECTION

Perhaps the most common and frequent method of quality control, superficial inspection, is the examination of the work surface to ensure effectiveness and workability. Surface inspection may be made with the naked eye or with a machinist's microscope.

TOLERANCE SYSTEM

The interchangeability of parts and products necessitates the need for minimum limits of perfection. These limits are established because it is impossible to make duplicate parts to the exact dimension specified. Factors such as machine variables and material composition will result in variations in the ultimate size of the finished product. However, it is possible to make component parts within varying dimensional limits and still allow them to fit and work together.

The amount of variation that is allowed is called *tolerance*. In other words, the part manufacturer will tolerate a certain amount of allowable error for that product. Therefore, when mechanics select a mass produced part, such as a 1/4-inch hex cap screw, they should be aware that its basic measurements will not be exact and will have some variations. For example the body diameter of this screw will range between 0.2500 and 0.2450 inch.

The amount of error that is allowed will be determined by the function that the part performs. If the part can have dimensional variation on two sides (i.e., plus and minus) it has *bilateral tolerance*. If the tolerance is specified in only one direction (i.e., either plus or minus), it has *unilateral tolerance*.

ALLOWANCE AND INTERFERENCE

In addition to allowing for the minimum and maximum sizes of parts, manufacturers also have to establish clearances between assembled parts. This is important in many units requiring an oil film to fit between moving parts for proper lubrication. Thus, when manufacturers make mating pieces, they will specify a specific clearance, known as *allowance*, between the parts.

Similar to tolerance, allowance may be expressed in terms of minimum (i.e., the tightest possible fit in assembly) or maximum (i.e., the loosest fit in assembly) allowances. The size of minimum and maximum allowances depends on the designed function of the mating parts. The mating parts cannot be so tight that they prevent lubrication, nor so loose that inefficiency results.

There are products that require two or more parts to be forced together into a more permanent assembly. When this is required, the external part will normally be of a constant size while the internal part will be oversized. Here, there will be no clearance space, but an opposition between parts. The amount of *opposition* is technically termed *interference*. Hence, interference is the negative allowance that exists between mating parts.

NONPRECISION AND PRECISION MEASUREMENT

To inspect and maintain machinery and equipment properly, the mechanic must be able to use measuring instruments. The use of measuring and testing tools, was covered to a great extent, in Chapter 2. However, there are several additional points that should be covered here.

Most measurement is classified according to the types of tools used for measuring—precision tools or nonprecision tools.

PRECISION INSTRUMENTS

Precision instruments or tools use complex mechanisms or devices to make measurements accurate to within thousandths of an inch or finer. Superprecision instruments are capable of measurements smaller than one ten-thousandth of an inch.

NONPRECISION INSTRUMENTS

Nonprecision instruments are simple and less accurate tools used for measuring to the line graduations on a rule. Examples of instruments that fall into this category are steel rules, the combination square, and calipers.

METRICATION

In 1975, the Metric Conversion Act was signed into law, which committed the United States to a conversion of all measurements to the metric system. The intent of the law was to bring the U.S.

to the same standard as most of the industrial nations of the world within ten years. Because compliance with the law is voluntary, the conversion has progressed at a much slower rate.

As a maintenance mechanic, you may be required to work with and maintain imported machinery that is built to metric standards. You will find it helpful to know some information about the metric system. Keep in mind that the metric system is a decimal (base ten) system of measurement. The standard for length is the meter, a unit that is slightly longer than a yard. While there are several subunits, you would mostly need only one—the millimeter (1,000 millimeters equal 1 meter). Metric drawings will use only the millimeter for length. Any metric architectural drawings will show only two units of length, meters and millimeters.

INTERNATIONAL SYSTEM OF UNITS

In 1954, the *Conference Générale des Poids et Mesures* adopted a standard metric system of measure that was based upon MKSA units: meter-kilogram-second-ampere. Later, the Kelvin, candela, and mole were added for temperature, luminous intensity, and substance quantity measures. This system of measurement was then named the *Système International d'Unités*, abbreviated SI.

METRIC EXPRESSIONS

There are specific procedures that are followed when working with metric measures. Multiples and submultiples of SI units are expressed in terms of prefixes, as listed in Table 11–1. Here, a unit of measure that is 1/100 of a meter is a *milli*meter (mm), while a unit that is 100 times larger than a meter is a *hecto*meter (hm).

TABLE 11–1

Prefixes for Unit Multiples and Submultiples

Prefix	Symbol	Unit Factor Multiple
tera	T	10^{12}
giga	G	10^{9}

TABLE 11–1 (*continued*)

Prefix	Symbol	Unit Factor Multiple
mega	M	10^6
kilo	k	10^3
hecto	h	10^2
deka	da	10
deci	d	10^{-1}
centi	c	10^{-2}
milli	m	10^{-3}
micro	μ	10^{-6}
nano	n	10^{-9}
pico	p	10^{-12}
femto	f	10^{-15}
atto	a	10^{-18}

The use of metric notations should follow standard procedures and rules. The first is that a space will always be placed between the numeric value and unit of measure. In other words, thirty-four meters is written as "34 m," not "34m." The only exception to this rule is temperature and angle measures. Here, the angle or degree symbol "°" is placed directly next to the numeric value—"212°", not "212 °."

When dealing with values less than 1 and greater than −1, it is recommended that a zero be placed in front of the decimal point, such as 0.234 and −0.897. This will help prevent misinterpreting the printing or writing of a faint or unclear decimal point. When writing decimals that exceed five digits, it is good practice to place a space after every third digit.

METRIC MEASURES

The only official length measure legalized by an act of Congress (1866) is the meter. By law, the legal equivalent of the meter is 39.37 inches. The official U.S. prototype meter, known as Meter No. 27, is preserved at the Bureau of Standards.

Traditional metric measures are normally broken down into six major groups: length, square, cubic, dry and liquid, and weight measures. The SI metric values of these are listed here.

LENGTH MEASURES

1 centimeter (cm) = 10 millimeters (mm)
1 decimeter (dm) = 10 cm
1 meter (m) = 10 dm
1 kilometer (km) = 1000 m

SQUARE MEASURES

1 square centimeter (cm^2) = 100 square millimeters (mm^2)
1 square decimeter (dm^2) = 100 square centimeters (cm^2)
1 square meter (m^2) = 100 square decimeters (dm^2)

SURVEYOR'S SQUARE MEASURES

1 are (a) = 100 square millimeters (mm^2)
1 hectare (ha) = 100 are
1 square killometer (km^2) = 100 hectares (ha)

CUBIC MEASURES

1 cubic centimeter (cm^3) = 1000 cubic millimeters (mm^3)
1 cubic decimeter (dm^3) = 1000 cubic centimeters (cm^3)
1 cubic meter (m^3) = 1000 cubic decimeters (dm^3)

LIQUID AND DRY MEASURES

1 centiliter (cl) = 10 milliliters (ml)
1 deciliter (dl) = 10 centiliters (cl)
1 liter (1) = 10 deciliters (dl)
1 hectoliter (hl) = 100 liters (l)

WEIGHT MEASURES

1 centigram (cg) = 10 milligrams (mg)
1 decigram (dg) = 10 centigrams (cg)
1 gram (g) = 10 decigram (dg)
1 dekagram (dag) = 10 grams (g)
1 hectogram (hg) = 10 dekagrams (dag)
1 kilogram (kg) = 10 hectograms (hg)
1 metric ton (t) = 1000 kilograms (kg)

METRIC CONVERSIONS

The conversion of metric measures to U.S. Customary measures, and vice versa, is a problem that may occasionally arise. Tables 11-3 through 11-8 provide conversion factors most helpful to individuals employed in the machine and maintenance mechanic field.

The use of these tables is quite simple. To convert from one system of measure to the other, all you have to do is multiply the original value by the conversion factor. For example, to convert 3 feet into centimeters, multiply 3 by 30.48 to find that 3 ft = 91.44 cm.

TABLE 11–3
Length Conversion Factors

	in	ft	yd	mile	mm	cm	m	km
1 in	1	0.0833	0.0278	...	25.40	2.540	0.0254	...
1 ft	12	1	0.333	...	304.8	30.48	0.3048	...
1 yd	36	3	1	...	914.4	91.44	0.9144	...
1 mile	...	5280	1760	1	...	...	1609.3	1.609
1 mm	0.0394	0.0033	...	...	1	0.100	0.001	...
1 cm	0.3937	0.0328	0.0109	...	10	1	0.01	...
1 m	...	3.281	1.094	...	1000	100	1	0.001
1 km	...	3281	1094	0.6214	...	...	1000	1

TABLE 11–4
Area Conversion Factors

	sq. in	sq. ft.	acre	sq. mile	sq. cm	sq. m
1 sq. in.	1	0.0069	...	...	6.452	...
1 sq. ft.	144	1	...	...	929.0	0.0929
1 acre	...	43,560	1	0.0016	...	4047
1 sq. mile	...	...	640	1	...	...
1 sq. cm	0.1550	...	...	...	1	0.0001
1 sq. m	1550	10.76	...	...	10,000	1

Table 11–5
Volume Conversion Factors

	cu. in.	cu. ft.	cu. yd.	cu. cm	cu. m	liter	US gal.
1 cu. in.	1	. . .	. . .	16.387	. . .	0.0164	. . .
1 cu. ft.	1728	1	0.0370	28,317	0.0283	28.32	7.481
1 cu. yd.	46,656	27	1	. . .	0.7646	764.5	202.0
1 cu. cm	0.0610	. . .	. . .	1	. . .	0.0010	. . .
1 cu. m	61,023	35.31	1.308	1,000,000	1	999.97	264.2
1 liter	61,025	0.0353	. . .	1,000.028	0.0010	1	0.2642
1 US gal.	231	0.1337	. . .	3785.4	. . .	3.785	1
1 Imp. gal.	277.4	0.1605	. . .	4546.1	. . .	4.546	1.201

TABLE 11–6
Weight Conversion Factors

	grain	oz.	lb.	ton	g	kg	metric ton
1 grain	1	. . .	. . .	. . .	0.0648	. . .	. . .
1 oz.	437.5	1	0.0625	. . .	28.35	0.0283	. . .
1 lb.	7000	16	1	0.0005	453.6	0.4536	
1 ton	. . .	32,000	2000	1	. . .	907.2	0.9072
1 g	15.43	0.0353	. . .	. . .	1	0.001	. . .
1 kg	. . .	35.27	2.205	. . .	1000	1	0.001
1 metric ton	. . .	35,274	2205	1.1023	. . .	1000	1

TABLE 11–7
Pressure Conversion Factors

	lb. per sq. in.	lb. per sq. ft.	int. atm.	kg per sq. cm	torr	in. Hg
1 lb/sq. in	1	144	. . .	0.0703	51.713	2.0359
1 lb/sq. ft.	0.00694	1	. . .	. . .	0.3591	0.01414
1 atmosphere (in'nat'l.)	14.696	2116.2	1	1.0333	760	29.921
1 kg/sq. cm	14.223	2048.1	0.9678	1	735.56	28.958
1 mm Hg (1 torr)	0.0193	2.785	. . .	. . .	1	0.0394
1 in. Hg	0.4912	70.73	0.0334	0.0345	25400	1

TABLE 11–8
Power Conversion Factors

	hp	watt	kw	Btu/ min.	Btu/ hr.	ft-lb/ sec.	ft-lb/ min.	metric hp.
1 hp	1	745.7	0.7475	42.41	2544.5	550	33,000	1.014
1 watt	. . .	1	0.001	0.0569	3.413	0.7376	44.25	0.00138
1 kw	1.3410	1000	1	56.88	3412.8	737.6	44.254	1.360
1 Btu/ min.	. . .	. . .	. . .	1	60	12.97	778.2	0.0239
1 metric hp	0.9863	735.5	0.7355	41.83	2509.8	542.5	32.550	1

PART FIVE

APPENDICES

Appendix 1

Job and Workplace Safety

Both journeymen and apprentice maintenance mechanics should be aware of those factors required for a safe and healthy work environment. Legal responsibilities are defined by the Occupational Safety and Health Administration (OSHA), whose objective it is to make all business, commercial, and industrial facilities safe and healthy. This chapter presents an overview of legal safety and health requirements in the workplace and specific guidelines and practices for work with various types of equipment, machinery, tools, and materials.

RESPONSIBILITIES UNDER OSHA

The Occupational Safety and Health Act (OSHA) of 1970 made safety and health on the job a matter of law for all businesses and their employees. OSHA functions to promote and implement safety and health standards, issue regulations, provide training programs for employers and employees, improve substandard working conditions, and assist in establishing plans and programs that will be in compliance with legal regulations.

To help carry out its legal requirements, OSHA has established ten regional offices throughout the United States. These regional offices administer and enforce, through on-site inspections, the safety and health standards specified by law. It is these regional offices that actually take action against noncomplying businesses.

PROBLEM-SOLVING ASSISTANCE

In many states, on-site consultation is available through OSHA. This activity is federally funded so that a business, with no fee or charge, may request a visit from a state consultant who can provide practical advice and recommendations for upgrading and improving job safety and health. These consultants have received the same training as federal and state inspection staffs, but they cannot issue citations, give penalties, or routinely provide information about your workplace conditions to inspectors. Their primary function is to help improve the work environment.

To request OSHA consultations, contact the nearest state office that provides this service. When agreement is reached to review your facilities, a four-stage process will be followed: 1) an opening conference, 2) a walk-through of the facilities, 3) a closing conference, and 4) a written summary of findings.

During the walk-through stage, the consultant will not only explain OSHA standards but will also identify those that apply to your situation. The consultant will explain the technical language of the standards, identify facilities and activities not in compliance, and when possible, suggest how they can be brought into compliance.

OTHER SOURCES OF HELP

In addition to OSHA consultation, there are other sources of help. Perhaps the most frequently used sources are workers' compensation carriers and other insurance companies. These companies often conduct periodic inspections and visits to assess the safety and health situations of their clients. Because services vary from one company to the next, each business should contact its own carrier to identify the available services.

Trade associations and employer groups can also be helpful. If you are a member, find out what the local organization does to assist members. If you are not a member find out whether or not these groups provide information to nonmembers (many do).

The National Safety Council (NSC) is a nonprofit organization that provides a broad range of information about workplace safety and health. Check to see whether a local chapter of the NSC exists in your area, and find out what materials are available that pertain to your situation. If no local chapter is nearby, write to the national office at:

National Safety Council
425 North Michigan Avenue
Chicago, IL 60611

FEDERAL STANDARDS

The law requires the U.S. Secretary of Labor to establish standards necessary to provide for a safe and healthy workplace. There are three separate sets of standards that every maintenance mechanic should be aware of:

1. General Industry (29 CFR 1910)
2. Construction (29 CFR 1926)
3. Maritime Employment (29 CFR 1915–1919)

Table AP1-1 provides a checklist that can help maintenance mechanics in a fact-finding inspection. This checklist is by no means all-inclusive. Each person is encouraged to add, delete, or modify each section as deemed necessary.

TABLE AP1–1
Self-Inspection Checklist[1]

Category	Condition	OK	Action Needed
Electrical Wiring, Fixtures, and Controls	Are workplace electricians familiar with the National Electrical Code® (NEC®)*?	☐	☐
	Do you specify compliance with NEC® for all contract electrical work?	☐	☐
	Do you have electrical installations in hazardous dust or vapor areas, and if so do they meet NEC® standards for hazardous locations?	☐	☐
	Are all electrical cords strung so they do not hang on pipes, nails, hooks, etc.?	☐	☐
	Are all conduit, BX cable, etc., properly attached to all supports and tightly connected to junction and outlet boxes?	☐	☐
	Is there evidence of fraying on any electrical cords?	☐	☐
	Are rubber cords kept free of grease, oil, and chemicals?	☐	☐

***National Electrical Code® and NEC® are registered trademarks of the National Fire Protection Association, Quincy, MA 02269.**

TABLE AP1–1 (*continued*)

Category	Condition	OK	Action Needed
	Are metallic cable and conduit systems properly grounded?	☐	☐
	Are portable electric tools and appliances grounded or of double insulated type?	☐	☐
	Are all ground connections clean and tight?	☐	☐
	Are fuses and circuit breakers the right type and size for the load on each circuit?	☐	☐
	Are all fuses free of "jumping" with pennies or metal strips?	☐	☐
	Do switches show evidence of overheating?	☐	☐
	Are switches mounted in clean, tightly closed metal boxes?	☐	☐
	Are motors clean and kept free of excessive grease and oil?	☐	☐
	Are motors properly maintained and provided with adequate over-current protection?	☐	☐
	Are bearings in good condition?	☐	☐

TABLE AP1–1 (*continued*)

Category	Condition	OK	Action Needed
	Are portable lights equipped with proper guards?	☐	☐
	Are all lamps kept free of combustible material?	☐	☐
	Is your electrical system checked periodically by someone competent in the NEC®?	☐	☐
Exits and Access	Are all exits visible and unobstructed?	☐	☐
	Are all exits marked with a readily visible sign that is properly illuminated?	☐	☐
	Are there sufficient exits to ensure prompt escape in case of emergency?	☐	☐
	Are controls in place for areas requiring limited occupancy?	☐	☐
	Do you take special precautions to protect employees during construction and repair operations?	☐	☐
Fire Protection	Are portable fire extinguishers provided in adequate number and type?	☐	☐

TABLE AP1–1 (*continued*)

Category	Condition	OK	Action Needed
	Are fire extinguishers inspected monthly for general condition and operability and dates noted on the inspection tag?	☐	☐
	Are fire extinguishers recharged regularly and properly noted on the inspection tag?	☐	☐
	Are fire extinguishers mounted in readily accessible locations?	☐	☐
	If you have interior stand pipes and valves, are these inspected regularly?	☐	☐
	If you have a fire alarm system, is it tested at least annually?	☐	☐
	Are plant employees periodically instructed in the use of extinguishers and fire protection procedures?	☐	☐
	If you have outside private fire hydrants, were they flushed within the last year and placed on a preventive maintenance schedule?	☐	☐

TABLE AP1–1 (*continued*)

Category	Condition	OK	Action Needed
	Are fire doors and shutters in good operating condition?	☐	☐
	Are fusible links in place?	☐	☐
	Are they unobstructed and protected against obstruction?	☐	☐
	Is your local fire department well acquainted with your facility, location, and specific hazards?	☐	☐
	Automatic Sprinklers:		
	Are water control valves, air and water pressures checked weekly?	☐	☐
	Are control valves locked open?	☐	☐
	Is maintenance of the system assigned to responsible persons or a sprinkler contractor?	☐	☐
	Are sprinkler heads protected by metal guards where exposed to mechanical damage?	☐	☐
	Is proper minimum clearance maintained below sprinkler heads?	☐	☐

TABLE AP1–1 (*continued*)

Category	Condition	OK	Action Needed
Housekeeping and General Work Environment	Is smoking permitted in "safe areas" only?	☐	☐
	Are NO SMOKING signs prominently posted for areas containing combustibles and flammables?	☐	☐
	Are covered metal waste cans used for oily and paint-soaked waste?	☐	☐
	Are rubbish and litter disposed of daily?	☐	☐
	Are paint spray booths, dip tanks, etc., and their exhaust ducts cleaned regularly?	☐	☐
	Have weeds or other combustible material been removed from within 20 feet of any buildings?	☐	☐
	For wet processes, are stand mats, platforms, or similar protection provided to protect employees from wet floors?	☐	☐
	Are waste receptacles provided for waste, and are they emptied regularly?	☐	☐

TABLE AP1–1 (*continued*)

Category	Condition	OK	Action Needed
	Do your toilet facilities meet the requirements of applicable sanitary codes?	☐	☐
	Are washing facilities provided?	☐	☐
	Are all areas of your business adequately illuminated?	☐	☐
	Are floor openings provided with toe boards and railings or a floor hole cover?	☐	☐
	Are stairways in good condition and standard railings provided for every flight having four or more risers?	☐	☐
	Are portable wood ladders and metal ladders adequate for their purpose, in good condition, and provided with secure footing?	☐	☐
	If you have fixed ladders, are they adequate, and are they in good condition and equipped with side rails or cages or special safety climbing devices, if required?	☐	☐

TABLE AP1–1 (*continued*)

Category	Condition	OK	Action Needed
	Are loading dock dockplates kept in serviceable condition and secured to prevent slipping?	☐	☐
	Is there means to prevent car or truck movement when dockplates are in place?	☐	☐
Machines and Equipment	Are all machines or operations that expose operators or other employees to rotating parts, pinch points or flying chips, particles, or sparks adequately guarded?	☐	☐
	Are mechanical power transmission belts and pinch points guarded?	☐	☐
	Is exposed power shafting less than 7 feet from the floor guarded?	☐	☐
	Are hand tools and other equipment regularly inspected for safe condition?	☐	☐
	Is compressed air used for cleaning less than 30 psi?	☐	☐
	Are power saws and similar equipment provided with safety guards?	☐	☐

TABLE AP1–1 (*continued*)

Category	Condition	OK	Action Needed
	Are grinding wheel tool rests set to within 1/8 inch or less of the wheel?	☐	☐
	Is there any system for inspecting small hand tools for burred ends, cracked handles, etc.?	☐	☐
	Are compressed gas cylinders regularly examined for obvious signs of defects, deep rusting, or leakage?	☐	☐
	Is care used in handling and storing of cylinders, safety valves, relief valves, etc., to prevent damage?	☐	☐
	Are all air receivers periodically examined, including the safety valves?	☐	☐
	Are safety valves tested regularly and frequently?	☐	☐
	Is there sufficient clearance from stoves, furnaces, etc., for stock, woodwork, or other combustible materials?	☐	☐

TABLE AP1–1 (*continued*)

Category	Condition	OK	Action Needed
	Is there clearance of at least 4 feet in front of heating equipment involving open flames as in gas radiant heaters, in front of firing door of stoves, furnaces, etc.?	☐	☐
	Are all oil and gas fire devices equipped with flame failure controls that will prevent flow of fuel if pilot or main burner are not working?	☐	☐
	Is there at least a 2-inch clearance between chimney brickwork and all woodwork or other combustible materials?	☐	☐
	For Welding or Flame Cutting:		
	Are only authorized, trained personnel permitted to use such equipment?	☐	☐
	Have operators been given a copy of operating instructions and been asked to follow them?	☐	☐
	Are welding gas cylinders stored so they are not subjected to damage?	☐	☐

TABLE AP1–1 (*continued*)

Category	Condition	OK	Action Needed
	Are valve protecting caps in place?	☐	☐
	Are all combustible materials near the operator covered with protective shields or otherwise protected?	☐	☐
	Is a fire extinguisher provided at the welding site?	☐	☐
	Do operators have the proper protective equipment?	☐	☐
Materials	Are approved safety cans or other acceptable containers used for handling and dispensing flammable liquids?	☐	☐
	Are all flammable liquids that are kept inside buildings stored in proper storage containers or cabinets?	☐	☐
	Do you meet safety standards for all spray painting or dip tank operations involving combustible materials?	☐	☐

TABLE AP1–1 (*continued*)

Category	Condition	OK	Action Needed
	Are oxidizing chemicals stored in segregated areas where they can have no contact with organic material except shipping bags?	☐	☐
	Do you have a NO SMOKING rule enforced in areas involving storage and use of hazardous materials?	☐	☐
	Are NO SMOKING signs posted where needed?	☐	☐
	Is ventilation equipment provided for removal of contaminants from such operations as production grinding, buffing, spray painting, and/or vapor degreasing, and is it operating properly?	☐	☐
	Are all personnel operating lift trucks properly trained?	☐	☐
	Is overhead protection provided on high lift rider trucks?	☐	☐

TABLE AP1–1 (*continued*)

Category	Condition	OK	Action Needed
	Are all materials used in your facilities checked for toxic qualities?	☐	☐
	Have control procedures been instituted for toxic materials, such as (where appropriate) respirators, ventilation systems, handling practices, etc.?	☐	☐

[1]Adopted from *OSHA Handbook for Small Businesses* Washington DC: U.S. Department of Labor, Occupational Safety and Health Administration, OSHA 2209, 1977.

Appendix 2

Codes, Regulations, and Standards

Modern businesses must have an on-going program of safety, health, fire, facility, machinery, and equipment checks, including plans for the development and specification of maintenance and servicing programs. Such programs not only save costly alterations and installations but also are local, state, and/or federal requirements. In fact, many standards and procedures used within the trade are derived directly from codes, regulations, and standards. This chapter presents an overview of codes, regulations, and standards relevant to the machine and maintenance mechanic.

LAW, STATUTE, ORDINANCE, AND CODE REQUIREMENTS

Codes and procedures relevant to facility and equipment maintenance vary among the fifty states, but they also have much in common. Most state governments have offices responsible for fire and safety protection, building and electrical codes, and mechanical codes. These offices function as central agencies for the sponsorship, promotion, and enforcement of related activities.

GOVERNMENTAL RESPONSIBILITIES

Although most code requirements are enforced at the local government level, others involve federal and state intervention. Therefore it is important that the machine and maintenance mechanic understand the role each governmental sector plays in this area.

FEDERAL GOVERNMENT

Federal laws and regulations deal primarily with businesses and individuals having a common relationship that extends beyond the boundaries of individual states, as in the field of transportation where numerous carriers cross state boundaries. To avoid the confusion that would arise from fifty different transport regulations, the Department of Transportation (DOT) and the Interstate Commerce Commission (ICC) have been given the responsibility for interstate and some local carrier safety.

The federal government also assumes responsibility for government-owned or occupied facilities, but in many cases it leaves the responsibility for inspection, fire suppression, and other activities to local governments. However, military bases and some other government installations are generally protected and inspected by federal government personnel.

As previously discussed, the federal government entered the area of employee safety with the enactment of the Occupational Safety and Health Act (OSHA) of 1970. However, the Act's programs are controlled by the states.

STATE GOVERNMENT

State code enforcement is generally associated with a specific state office, such as the state fire marshall's office. Most state laws concern public safety in assemblage occupancies, institutions such as schools and hospitals, and regulations dealing with combustible and flammable liquids and gasses.

LOCAL GOVERNMENT

The local government defined as county and/or municipal (town or city), concerns itself with the remaining areas of code and regulatory enforcement. Most local governments have laws that overlap into areas addressed by the state and federal government. In many cases, local requirements are more stringent than state and federal codes.

LAWS, STATUTES, ORDINANCES, AND CODES

Building, fire, equipment, and machine regulations and standards are regulated at different levels of government. Laws, statutes, ordinances, and codes make up the foundation upon which these programs are built. They are enacted to define the limits of action and impose restrictions on unsafe and hazardous situations.

A *law* is a system of rules prescribed under authority of a level of government. *Statutes* are laws that come from the state level; *ordinances* are provisions enacted by a municipal government for local application. A *code* is a body of laws systematically specified for easy reference.

Local codes generally introduce restrictions for the machine and maintenance mechanic. The restrictions are designed to promote the general safety and health of the public.

There are two general categories of code documents used in the United States.

1. *Performance Codes.* These codes set standards for the industry or activity. An example of this type of code is the Uniform Building Code, which was written by the International Conference of Building Officials.
2. *Specification Codes.* These codes specify how the performance code objectives are to be met. An example of a specification code is the National Building Code recommended by the American Insurance Association.

The National Building Code was first published in 1905 and served as a *model code.* Model codes have no legal status, but provide a guide for state and local jurisdictions in enacting their own codes. Other examples of model codes are the National Electrical Code, Standard Fire Prevention Code, and Uniform Mechanical Code.

MODEL CODES

Though model codes have no legal status, it is important for both the apprentice and journeyman to know those that apply to their situation. Mechanics should also make themselves familiar with the state and municipal legal codes pertaining to their job responsibilities.

UNIFORM MECHANICAL CODE

The Uniform Mechanical Code, published by the International Conference of Building Officials, is sponsored jointly by the International Association of Plumbing and Mechanical Officials and the International Conference of Building Officials. The code was developed for the installation of mechanical systems, their alteration, repair, and replacement. It includes information pertaining to appliances, fixtures, and fittings such as ventilating, air conditioning, incinerating equipment, and other energy-related systems.

The major sections and content of the uniform mechanical code are described here.

Program Administration. This section includes the scope and organization of the program, permits and fee structures, inspection procedures, certificates of approval, violations and penalties, appeal processes, and validity of the program.

Definitions and Standards. This section provides technical definitions and standards for terms included in the code.

Air Conditioning, Heating, and Ventilation Equipment. This section covers the design, construction, installation and repair of equipment used in related systems. Equipment specifically covered includes air conditioning equipment, heating equipment, combustion and ventilation of air, chimneys and vents, ventilation systems, and exhaust systems.

Refrigeration. The classification of refrigeration systems and refrigerants, the proper location of equipment, installation requirements, and field testing are included in this part.

Ducts and duct systems. This area deals with all ductwork used in air conditioning, heating, ventilating, exhaust, and conveying systems. Information is provided about duct construction, flexible connectors, vibration isolation connectors, insulation, ducts in concrete slabs, supports and hangers, fire protection of ducts, air filters, weatherproofing, plenum chambers and floor systems, and fresh air intakes.

Piping standards. This section addresses factors such as piping used for steam and hot water, condensate drain, chilled water, condenser water, make-up water, fuel piping, refrigerant, and testing and inspection procedures.

Incinerator Requirements. This part of the code defines general specifications for new and altered installations and for approved equipment as well as information pertaining to locations and clearances, classification of incinerators and waste, gas-fired incinerators, gas burners, scrubber or gas washer, stacks and chimneys, spark arrestors, marking incinerators, and proper operation of incinerators.

Electrical Requirements. This last section of the Uniform Mechanical Code includes information about electrical connections, ignition and control devices, grounding, circuitry, and continuous power.

STANDARD FIRE PREVENTION CODE

All Standard Fire Prevention Codes are based on the model fire code that was developed by the National Fire Protection Association. This code provides specific regulations based on recognized practices for the protection of life and property from the hazards of fire. The enforcement of these laws is the responsibility of local fire officials, such as the municipal fire chief. Most large industrial complexes have their own fire brigade and fire safety engineer. There are numerous topics covered in fire prevention codes, all relating to the proper handling of flammable material and hazardous situations. Examples of typical code topics are briefly described here.

Administrative Provisions. These provisions address a number of management concerns including the purpose of the code, appeals and adjustments, recognition of standards and publications used in the code, required permits and certificates, and definitions of technical terms.

Facility Requirements. Specific requirements for particular industries, such as automobile tire rebuilding plants, waste handling concerns, bowling establishments, dry cleaning operations, and motion picture businesses, are covered under this topic.

Handling and Use of Combustible Materials. This section deals with materials such as cellulose nitrate and compressed gases.

Fire Prevention for the Application of Flammable and Combustible Materials. This topic deals with various industrial processes, such as spray finishing, dip tanks, electrostatic apparatus, automobile undercoating, powder coatings, organic peroxides and dual component coatings, tank and bulk storage, materials handling, and processing liquids, and the fire prevention techniques applicable to these processes. Also, guidelines for refineries, chemical plants, distilleries, transporting pipelines, crude oil production, and tank vehicles used for flammable and combustible materials are provided.

Fireworks and Explosives. This section includes the handling, transporting, and storage of all explosives, blasting agents, and ammunitions. The prevention of dust explosions is also considered.

Processing Materials. This topic covers processing materials including those used in fruit ripening processes, fumigation and thermal insecticidal fogging, liquified petroleum gases, and hazardous chemicals. In addition, regulations are provided for garages, lumber yards and woodworking plants, ovens, industrial baking and drying, and for places of assembly.

General Fire Precautions. This general topic addresses open burning and incinerators, flammable and combustible materials, fire reporting and false alarms, use of equipment and devices, and vacant buildings.

Special Situations. Various types of special situations, including tents and air supported structures, mechanical refrigeration, welding and cutting calcium carbide and acetylene, working with magnesium, the manufacturing of organic coatings, high piled combustible stock, and motion picture projection, are covered under this topic.

Cryogenic Fluids. This part covers containers, storage, piping, process and accessory equipment at manufacturing facilities and consumer sites, handling, loading, and unloading, and transportation.

Airports, Heliports, and Helistops. This section provides regulations about air operation, refueler units, and helistops.

NATIONAL ELECTRICAL CODE®

The first National Electrical Code® was developed in 1897 because of the efforts of insurance, electrical, architectural, and related interests. Since 1911, the code has been sponsored by the National Fire Protection Association. The code is periodically amended to address new concerns about equipment, materials, components, and procedures.

The major points addressed in the National Electrical Code® are briefly described here.

General Information. This part of the code addresses definitions and requirements for electrical installations.

Wiring Design and Protection. This section covers the use and identification of grounded conductors, branch circuits, feeders, branch circuit and feeders calculations, outside branch circuits and feeders, services, overcurrent protection, grounding, and surge arresters.

Wiring Methods and Materials. The wiring methods and materials section is the most extensive section of the code and includes points on wiring methods, temporary wiring, cable trays, medium voltage cable, shielded nonmetallic sheathed cable, underplaster extensions, rigid metal and nonmetallic conduits, flexible metallic tubing and metal conduit, surface raceways, cellular metal floor raceways, busways, cabinets and cutout boxes, switches, and switchboards and panelboards.

Equipment for General Use. This topic addresses practices for flexible cords and cables, fixture wires, lighting fixtures and related components, appliances, fixed electric space heating equipment, fixed outdoor electric de-icing and snow melting equipment, fixed electric heating equipment for pipelines and vessels, motors, motor circuits and controllers, air conditioning and refrigeration equipment, generators, transformers and transformer vaults, capacitors, resistors and reactors, and storage batteries.

Special Appliances. This section covers the proper placement of different classes of materials, hazardous locations, commercial garages used for repair and storage, aircraft hangers, gasoline dispensing service stations, bulk storage plants, finishing processes, health care facilities, places of assembly, theaters, motion picture and television studios, motion picture projectors, manufactured buildings, agricultural buildings, mobile homes and parks, recreational vehicles and parks, and marinas and boatyards.

Special Equipment. Topics dealt with here include electric signs and outline lighting, manufactured wiring systems, cranes and hoists, elevators and related equipment, sound recording equipment, data processing systems, organs, X-ray equipment, induction and dielectric heating equipment, electrolytic cells, electroplating, metalworking machine tools, electrically driven or controlled irrigation machines, swimming pools and fountains, and integrated electrical systems.

Special Conditions. This section addresses emergency systems, legally required standby systems, optional standby systems, over 600-volt nominal circuits and equipment, circuits and equipment operating at less than 50 volts, various classes of remote-control and signalling power limited circuits, and fire protective signaling systems.

Communications Systems. Recommended standards and practices for communication circuits, radio and television equipment, and community antenna television and radio distribution systems are covered under this code heading.

Tables and Examples. This part includes sample problems and how they should be solved, as well as reference tables pertaining to dimensions, sizes, and rating criteria of various materials and components.

UNIFORM BUILDING CODE

The Uniform Building Code covers the fire, life, and structural safety aspects of all buildings and related structures. Similar to the Uniform Mechanical Code, this code is also published by the International Conference of Building Officials. The major topics covered in this code are described here.

Program Administration. This topic describes the title and scope of the code, its organization and enforcement, and required permits and inspections.

Definitions and Abbreviations. Definitions of technical terms used throughout the code as well as abbreviations are included in this section.

Requirements Based on Occupancy. This section provides a classification of all buildings by use or occupancy and general requirements for all occupancies. Topics addressed per classification include construction, property location, exit facilities, lighting, ventilation, sanitation, shaft enclosures, fire extinguishing systems, special hazards, and parking areas.

Requirements Based on Location in Fire Zones. This topic identifies restrictions for Fire Zones No. 1 through 3.

Requirements Based on Types of Construction. This section classifies types of construction and general requirements for each and also provides recommended standards for types I, II, III, IV, and V fire resistive buildings.

Engineering Regulations. This part describes the quality and design of construction materials. This includes general design requirements, masonry, wood, concrete, steel, and aluminum materials.

Detailed Regulations. This section provides detailed rules for excavations, foundations and retaining walls; veneer, roof construction and covering; stairs, exits and occupant loads; skylights, sound transmission control, penthouse and roof structures; masonry or concrete chimneys, fireplaces and barbecues; fire extinguishing systems, stages and platforms; and motion picture projection rooms.

Fire Resistive Standards for Fire Protection. This part provides guidelines for interior wall and ceiling finish and other fire resistive standards.

Regulations for Use of Public Streets and Projections Over Public Property. This topic addresses the protection of pedestrians during construction or demolition and during the permanent occupancy of public property.

Wall and Ceiling Coverings. This section details recommendations for wall and ceiling coverings with particular attention to topics such as materials, vertical and horizontal assemblies, interior and exterior lath and plaster, exposed aggregate plaster, gunite, gypsum wallboard, softwood plywood paneling, and shear-resisting construction with wood frame.

Special Subjects. This topic includes film storage, prefabricated construction, elevators and other moving equipment, plastics, and glass and glazing.

Legislative Topics. This section describes the validity of legislation, uniform building code standards, and ordinances as well as repeal procedures and effective dates.

Appendices. Included among the appendices are standards and tables for weights of building materials, sound transmission control, fire extinguishing systems, photographic and X-ray films, patio covers, fallout shelters, excavation and grading, and other specific topics.

SOURCES OF INFORMATION AND AID

Machine and maintenance mechanics often require special up-to-date information that is not readily available. To help the apprentice and tradesman, certain trade associations, public agencies, and private service organizations publish informative bulletins. The number of organizations and their total publications is too great to include in this book. However, there are several sources of information that the apprentice and mechanic should be aware of.

SERVICE ORGANIZATIONS

NATIONAL SAFETY COUNCIL

One of the best sources for information about work place safety and the prevention of injury is the National Safety Council (NSC). Its main office is located at 425 N. Michigan Avenue, Chicago, Illinois 60611. The Council also provides statistical information in annual and monthly publications, technical materials, and training materials, all of which are listed in the current NSC catalog.

INDUSTRIAL HEALTH FOUNDATION

One other service organization that the mechanic should be aware of is the Industrial Health Foundation, Inc. Headquartered at 5231 Centre Avenue, Pittsburgh, Pennsylvania 15232, the foundation is a nonprofit research organization of industries and advocates of industrial health programs that aim to improve working conditions and better human relations.

The foundation is made up of a technical staff who provide direct professional assistance to members in the study of industrial health hazards and their control; help companies develop health programs as part of their organization; and provide and promote technical advancement of industrial medicine and hygiene by investigation, research, and other activities.

Their activities are classified into the following areas: medicine, chemistry, toxicology, industrial hygiene, engineering, research, and law.

STANDARDIZATION GROUPS

Throughout this book several standards organizations have been mentioned and their guidelines used.

ANSI

The American National Standards Institute (ANSI), headquartered at 1430 Broadway, New York, New York 10018, is the umbrella organization for many groups. It also represents the United States in international standards development through the International Organization for Standardization (ISO), the International Electrotechnical Commission (IEC), and the Pacific Area Standards Congress (PASC).

ASTM

The world's largest source of voluntary consensus standards for materials, products, systems, and services is the American Society for Testing and Materials (ASTM) at 1916 Race Street, Philadelphia, Pennsylvania 19103. The ASTM draws its membership from a broad area of industries, agencies, and individuals concerned with materials.

OTHER CODE/STANDARDS GROUPS

In addition to ASTM and ANSI there are many industries that provide codes and standards through their trade associations. Some of these groups and their addresses are:

American Gas Association, Inc. Laboratories
8510 East Pleasant Valley Road
Cleveland, Ohio, 44131

American Society of Heating, Refrigeration,
and Air Conditioning Engineers
United Engineering Center
345 East 47th Street
New York, New York 10017

American Society of Mechanical Engineers
United Engineering Center
345 East 47th Street
New York, New York 10017

American Society of Sanitary Engineering
4328 South Western Avenue
Chicago, Illinois 60609

American Welding Society
2501 NW 7th Street
Miami, Florida 33125

Cast Iron Soil Pipe Institute
1824-26 Jefferson Place, NW
Washington, DC 20036

Committee of Steel Pipe Producers
American Iron and Steel Institute
150 East 42nd Street
New York, New York 10017

Compressed Gas Association
500 Fifth Avenue
New York, New York 10036

Federal Specification
Superintendent of Documents
Government Printing Office
Washington, DC 20402

Incinerator Institute of America
60 East 42nd Street
New York, New York 10017

International Conference of Building Officials
5360 South Workman Mill Road
Whittier, California 90601

Manufacturers Standardization Society of the
Valve and Fittings Industry
420 Lexington Avenue
New York, New York 10017

National Oil Fuel Institute, Inc.
60 East 42nd Street
New York, New York 10017

National Plant Food Institute
1700 K Street, NW
Washington, DC 20006

Plastic Pipe Institute
A Division of the Society of the Plastic Industry, Inc.
250 Park Avenue
New York, New York 10017

Sheet Metal & Air Conditioning Contractors
National Association, Inc.
1611 North Kent Street
Arlington, Virginia 22209

Underwriters' Laboratories, Inc.
207 East Ohio Street
Chicago, Illinois 60611

Uniform Boiler and Pressure Valve Laws Society, Inc.
57 Pratt Street
Hartford, Connecticut 06103

Glossary

A

ANSI. American National Standards Institute.

ASME. American Society of Mechanical Engineers.

Aftercooler. *See* heat exchanger.

Air Conditioning. Cooling and heating systems.

Ammeter. An electrical instrument used to measure electrical current flow.

Amp. The unit of measurement for electrical current.

Arc Welding. Welding processes employing electrical arcs between the part and an electrode to generate heat.

Assembly Drawings. Drawings showing multiple-part products as they appear totally assembled.

Auger Bit. Cutting tools used to bore holes in wood and other soft materials. *See* bit.

Axonometric Drawing. A pictorial drawing projection technique such as isometric and dimetric.

B

Barometer. An instrument used to measure atmospheric pressure.

Basic Size. The size from which the limits of size are determined by using allowances and tolerances.

Batteries. Devices used to store electrical energy.

Bearings. Devices used to support rotating or oscillating parts and protect them from damage.

Bit. Cutting tools for drilling or boring holes.

Bit Gage. A drill bit attachment used for boring holes to a specific depth.

Blueprint. A reproduced drawing. Traditionally it appeared as white lines with deep blue background. Today prints usually appear as blue-lined drawings on a white background.

Boiler. A pressure vessel generating steam for heating and power.

Bourdon-Tube Gauge. An instrument used to measure positive and negative pressures.

Brace. A tool used to hold bits for drilling or boring holes.

Brazing. The joining of metals with a filler metal at temperatures above 800° F; sometimes referred to as hard soldering.

Bus. Electrical line supplying power to machinery, equipment, or instruments.

C

Caliper. Used for transfering measurements from one location to another; a layout tool.

Capillary Attraction. The property of liquid to flow.

Centrifugal Pumps. Equipment used for moving fluid.

Chain Dimensioning. A dimensioning technique where each dimension starts and finishes with a preceding dimension.

Chisel. A shaving tool used to remove material by shearing, cutting, and/or chipping.

Clearance. The difference in sizes between mating parts where the female part is larger than the male part.

Cloud Point. The temperature of lubricants where there is a separation of wax or hydrocarbon components; also known as FLOC point for oil.

Compressor. Equipment used to convert the mechanical energy transmitted by a motor into compressed air and gas.

Constant Current Charging. The charging method used for many nickel-cadmium-alkali batteries.

Constant Voltage Charging. The charging method used for most nickel-cadmium-alkali batteries.

Contactor. Electrical devices used to handle high voltages and currents.

Control (electrical). Devices used to control the flow of electrical power for the proper operation of machinery, equipment, and instruments.

Coordinate Axis. Directional measures and/or movements used in drawings and machinery, consisting of three major axes: x, y, and z.

Coordinate Dimensioning. A dimensioning technique where measurements are taken from coordinate axes.

Corner Brace. A type of brace used for boring operations in corners and against walls.

Corrosion. The destructive chemical or electrochemical attack on materials.

Coulomb. The unit of measure for electrical charge.

Curved Claw Hammer. A common tool used for carpentry work.

D

Deferred Concrete Floors. Floors applied to a preset base slab.

Densification. The process of correcting and strengthening monolithic floors.

Detailed Drawings. Orthographic drawings of a single part or product.

Divider. A tool used to lay out regular curves and transfer dimensions.

Drills. A classification of boring tools that are either manually or power driven.

Drawfiling. An operation by which a file is moved in short strokes and positioned perpendicular to the work.

Dropping Point. The property of greases where it changes from its thickened state to that of a liquid at a given temperature.

Drywall Hammer. A hammer used for installing drywall.

E

Exponent. A mathematical note expressed as a superscript, sometimes referred to as a power note.

Extreme-Pressure (EP) Lubricants. Specialized lubricants capable of working under severe conditions and pressures to prevent material deterioration and corrosion.

F

Farad. The unit of measure for electric capacitance.

File. A finishing or semifinishing tool used for shaving material.

Filler Metal. Used in welding operations to provide metal to a welded joint; usually is provided by a welding rod.

Fits. The degree and quality of closeness with which the surfaces of parts are brought together.

Flooring Hammer. A hammer used for laying tongue and groove hardwood floors.

Fluid Power. The use of liquids and gases in the form of a fluid to produce power.

Force Fit. Also known as a shrink fit, characterized by a constant pressure on mating parts. *See* interference.

Friction Clutch. Machine element used to enable one rotating element to rotate and transmit torque to another element by means of friction.

G

Galvanic Corrosion. The corrosion caused when two dissimilar metals are brought into contact with each other.

Galvanometer. An electrical instrument used for measuring direct current generating devices.

Gas Welding. Welding processes employing a gaseous mixture as the fuel for the heat source.

Gage. Any instrument against which measurement is compared to a given standard or system, such as angle gage blocks.

Grinding. The removal of material by an abrading action.

Gun Tacker. A mechanical stapler.

H

Heat Exchanger. Equipment used to subcool liquids and cool compressor discharge air and gas.

Henry. The unit of measure for inductance.

Hydraulic Systems. A segment of fluid mechanics that employs the use of oil and other liquids.

I

ISO. International Organization for Standardization.

Interference. The difference is sizes between mating parts where the male is larger than the female part.

Isometric Drawing. An axometric projection drawing where the edges of the object project back at 30° angles, and form three equal 120° angles.

J

Joule. Unit of measure for work, energy, and quantity of heat.

L

Limits. The absolute minimum and maximum sizes of a part.

Level. A tool used to level members of a building, machine, or piece of equipment.

Location Fits. Used for determining the location of mating parts.

Lubricant. Any material that reduces friction and prevents material wear.

Lumen. The unit of measure for lumous flux.

Lux. The unit of measure for illumination.

M

Machinist's File. A category of files used for work on metal and other hard materials.

Mallet. A hammering tool with a soft head that is used for driving other tools.

Micrometer. A group of measuring instruments capable of measurements to within 0.001 inch, or 0.0001 inch with a vernier.

Milling. A common metal machining operation used to produce surface contours by a rotating cutter.

Monolithic Floors. Floors made up entirely of ready-mix concrete having excess water and preventing it from reaching maximum strength.

N

Newton. Unit of measure for force.

Noble Metal. Metals with more oxide.

O

OSHA. Occupational Safety and Health Administration.

Oblique Drawing. A pictorial drawing where one face of an object appears as an orthographic view, and the other sides project back at a standard angle.

Ohm. The unit of measure for electrical resistance.

Ohm's Law. A law stating the relationship between electrical voltage (E), current (I), and resistance (R) where $E = IR$.

Orthographic Projection. System of drawing used for preparing technical, engineering, and other types of working drawings; *see* principle views of projection.

Oxidation. A form of corrosion resulting in the formation of oxides; normally occurs at elevated temperatures.

P

PVC. Polyvinyl chloride, material that is commonly used for pipes and tubing.

Paint. A protective and/or decorative covering made up of pigments or solid powders, and binders or vehicles.

Parallel Dimensioning. A dimensioning technique in which all dimensions have a common reference point from which all measures are taken.

Peen Hammer. A hammer used to indent or compress metals.

Pitot Tube. A device used to measure static, total, and velocity pressure of a gas or a liquid.

Piting. The electrochemical corrosion at anodic or positively charged areas.

Plane. A wood shaving tool commonly used in carpentry work.

Pliers. Gripping tools.

Plumb Bob. Used in combination with a level to assure that a component is vertical.

Pneumatic System. A segment of fluid power based upon the use of air flow for power transmission and controlling devices.

Pour Point. The temperature at which oils will just begin to flow.

Precipitation Number. Used for steam cylinder and black oils to indicate the amount of carbonaceous matter and formed asphaltenes.

Primary Views. Based upon the principles of orthographic projection; consists of the front, top, and right side view of an object.

Principle Views of Projection. The principle views from which an object can be viewed by orthographic projection techniques. Consists of the front, top, right side, left side, bottom, and rear views of an object.

Profiling Instruments. Measuring instruments used to check degree of roundness.

Punch. Used for layout work, and is of two common types. Center punches are used to locate the center of holes for drilling, and prick punches are used for scribing on the workpiece.

R

Ratchet. A tool feature that allows a tool to turn in one direction, while it can turn in either clockwise or counter-clockwise direction.

Rheostat. A device used for starting and controlling the speed of dc and ac motors.

Rule. A basic measuring tool.

Running and Sliding Fits. Limits of clearance for running performance.

S

SAE. Society of Automotive Engineers.

SI. Système International d'Unités, a system of metric measurement.

SUS. Saybolt Universal Seconds. Units used to express the viscosity of lubricants.

Saw. A broad category of tooling used to cut soft and hard materials. Is either manual or powered.

Scientific Notation. Mathematical notation by powers of ten used to indicate the location of the decimal point.

Screwdriver. Tool used for driving screw materials. Is available in different point designs and sizes.

Shutdown Operations. A preplanned maintenance procedure where the entire facility is closed for a period of time for the purpose of performing specific maintenance operations.

Sine Bar and Plate. Tools used for providing angular measures by use of standard angles.

Solder. A lead-tin alloy used in soldering operations.

Soldering. A metal joining process employing filler metals at temperatures below 800°F.

Snips. A sheet material cutting tool that can be either a hand- or bench-type.

Square. A layout tool with two edges that are perpendicular to each other.

Straight Filing. An operation where a file is pushed along the length of its blade either parallel or at a slight angle to the workpiece.

Strain Gage. A device using electrical resistance change to measure strain.

Surface Texture. The degree and type of surface smoothness and roughness.

Switchgear. Electrical devices used for power distribution (circuit breakers and relays).

T

Tachometer. A velocity-measuring instrument.

Thermostat. An electrical instrument used to control one or more sources of heating and cooling.

Tolerance. The total amount of size variation that is allowed.

Transformers. Electrical equipment used to step-up or step-down electricity in a circuit.

Turbine. Machines used for generating rotary mechanical power from the energy of steam.

Turning. A machining operation employing a lathe and cutting tools.

Twist Drill. A bit used to bore holes that is designed for drilling into a wide variety of materials.

Two-Rate Method. The charging method used for lead-acid batteries incorporating a high voltage charge at the beginning and lower charge at the end of the charging cycle.

V

Venturi Tube. A device used to measure average flow rates.

Vernier Instruments. Measuring instruments based upon the relationship between two different graduated scales.

Viscosity. The fluidity of fluids expressed as the resistance to flow.

Vitrified. Type of bonding used in grinding wheels.

Volt. The unit of measure for electromotive force (emf) or electric potential.

Voltmeter. An electrical instrument used to measure voltage within a circuit. *See* volt.

W

Watt. The unit of measure for power.

Weber. The unit of measure for magnetic flux.

Welding. The fusing of material by employing high temperatures. *See* gas welding and arc welding.

INDEX